D1059887

A FIELD GUIDE TO
BACKYARD
BIRDS
OF NORTH AMERICA

Inspiring | Educating | Creating | Entertaining

Brimming with creative inspiration, how-to projects, and useful information to enrich your everyday life, quarto.com is a favorite destination for those pursuing their interests and passions.

© 2007, 2022 Quarto Publishing plc

This edition published in 2022 by Chartwell Books,
an imprint of The Quarto Group
142 West 36th Street, 4th Floor
New York, NY 10018 USA
T (212) 779-4972 F (212) 779-6058
www.Quarto.com

Contains content originally published as *Backyard Bird Tracker* in 2007

All rights reserved. No part of this book may be reproduced in any form without written permission of the copyright owners. All images in this book have been reproduced with the knowledge and prior consent of the artists concerned, and no responsibility is accepted by producer, publisher, or printer for any infringement of copyright or otherwise, arising from the contents of this publication. Every effort has been made to ensure that credits accurately comply with information supplied. We apologize for any inaccuracies that may have occurred and will resolve inaccurate or missing information in a subsequent reprinting of the book.

10 9 8 7 6 5 4 3 2 1

Chartwell titles are also available at discount for retail, wholesale, promotional, and bulk purchase. For details, contact the Special Sales Manager by email at specialsales@quarto.com or by mail at The Quarto Group, Attn: Special Sales Manager, 100 Cummings Center Suite 265D, Beverly, MA 01915, USA.

ISBN: 978-0-7858-4075-6

Conceived, edited, and designed by Quarto Publishing,
an imprint of The Quarto Group
The Old Brewery
6 Blundell Street
London N7 9BH UK

Assistant Editor: Charlene Fernandes
Editorial Assistant: Ella Whiting
Junior Designer: India Minter
Designer: Sally Bond
Art Director: Gemma Wilson
Sales Director: Nikki Tilbury
Publisher: Lorraine Dickey

Printed in China

A FIELD GUIDE TO
BACKYARD
BIRDS
OF NORTH AMERICA

A VISUAL DIRECTORY OF THE MOST
POPULAR BACKYARD BIRDS

ROB HUME

chartwell
books

Contents

Introduction

What is it that draws us to birds? Why is it that watching birds is one of the fastest growing hobbies in the world? There is no single answer, but rather a multitude of answers that are all focused on the bird-human interface: most birds are active when we are active.

Their language to us is the language of music—song, percussion, rhythmic, soothing, excited, incredibly varied, yet predictable so that we can identify birds by their voices as we would identify an old friend on the telephone. Birds are often colorful, with each species sporting unique hues and patterns of plumage that allow us to identify it as we might recognize an acquaintance on a busy street.

Our fine-feathered friends

Birds remind us of family and friends by behaviors ranging from the protective attention of parent birds toward their young, to the rambunctious "Feed me! Feed me!" demands of a fledgling, to the bill held high in a face-off of two birds in a territorial dispute at a mutual boundary, to the intimate collaboration of a pair constructing a nest, to the perceived joy in the songs of birds proclaiming the arrival of spring and the beginning of a new cycle of life. In short we are drawn to birds by their "human" qualities—which, of course, are not human at all, but the perception of our inquisitive minds.

Perhaps we are drawn to birds out of envy—that ability to fly, to soar above the clouds? Or perhaps it is merely the wonder we feel on observing the aerobatics of swallows, the speed of falcons, the effortless soaring of an eagle, the precision flight of a skein of geese or a wheeling flock of hundreds of blackbirds that seems to pulse as if the flock itself were a living being.

We are also drawn to birds by hope. Perhaps it is the hope that they might take advantage of the well-stocked feeder or birdbath we so carefully selected and placed in the garden. Maybe it is the dream that a new species will appear in our yard or that we might find a particular rare bird on a trip to its reported haunts. Or maybe it is the hope we have with

MALLARDS

These ducks rarely dive into the water for food, preferring to grab morsels on the water's surface.

the warming days of spring that the birds that nested nearby last year will return.

Perhaps it is the sense of satisfaction, pride, fulfillment, and maybe even the power that we might feel when bluebirds take up residence in the birdhouse we just built and carefully located in our backyard. Perhaps it is our amazement at their arrival back in our yards after hard winters and long migrations. How could we not respect these tiny feathered creatures that dauntlessly withstand snow and storms, heat and drought, and then saucily proclaim their survival with beautiful song. Perhaps it is seeing the awe on a child's face when peering for the first time into a bird nest with eggs or nestlings.

Birds share our lives by contributing to our aesthetic senses and emotional well-being, but they are also, in a sense, our "protectors," consuming insect pests and weed seeds and serving in many ways as barometers of the health of the ecosystems in which we live. Yes, they can be a nuisance. Blackbirds and European Starlings can consume crops and sometimes monopolize our feeders, but they also consume cutworms and other harmful pests. All too often with such birds, we think only of the negative. Proper accounting demands that we look at both sides of the ledger. And we should take a closer look at the ledger of our relationships to birds. We might provide feed and shelter for birds in our yards, but how much of their habitat have we destroyed?

A new relationship

This book is a primer for understanding our relationship with birds and their relationship with us. It provides a window on the intimate lives of birds and how they cope with the world in which we both live. Paired with a field guide and a pair of binoculars, this book can open

GREAT SPOTTED WOODPECKER

Its "drumming," usually on dead wood, can be heard up to half a mile away.

your eyes to a wonderful world that has been there all along, but which so often has gone "unseen." Only once you know what you are looking at, can you experience it fully. It is the beginning of a beautiful friendship.

The final section of this book provides a log for you to keep track of your explorations and observations of the world of birds. It is a repository for memories, but can also be a well from which you can draw understanding. Record the comings and goings of birds. When did they arrive in spring? When did they build their nest? How many young did they raise? What were they eating? How did they interact with other birds? While we know a great deal about birds, there is much we don't know. Share your discoveries and learn from others. Get involved with a local bird club, Audubon chapter, or regional ornithological society. The wonder of birds and excitement of birding is yours for life.

Jerome A. Jackson (Professor of Biology) and Bette J. S. Jackson (Associate Professor of Biology) Florida Gulf Coast University

Bird finder at a glance

Here is an overview of the 52 birds that feature in the US Backyard Bird Catalog. Turn to the relevant pages to learn more about their characteristics and habitats.

65
American Kestrel

65
Ring-billed Gull

66
Rock Pigeon

69
Chimney Swift

69
Ruby-throated
Hummingbird

70
Red-bellied Woodpecker

70
Northern Flicker

74
Purple Martin

75
Blue Jay

75
California Scrub-Jay

62
Double-crested Cormorant

63
Mallard

63
American Coot

64
Sharp-shinned Hawk

66
Mourning Dove

67
Yellow-billed Cuckoo

68
Great Horned Owl

71
Red-headed Woodpecker

72
Hairy Woodpecker

73
Tree Swallow

73
Barn Swallow

76
American Crow

76
Black-capped Chickadee

77
Oak Titmouse

78
White-breasted Nuthatch

79
House Wren

79
Winter Wren

80
Golden-crowned Kinglet

80
Ruby-crowned Kinglet

83
American Robin

84
Loggerhead Shrike

84
Cedar Waxwing

88
White-crowned Sparrow

89
House Sparrow

89
Brown-headed Cowbird

92
Pine Siskin

92
American Goldfinch

93
Purple Finch

93
House Finch

81
American Redstart

82
Wrentit

83
Yellow-rumped Warbler

85
European Starling

86
Fox sparrow

86
Song sparrow

87
Dark-eyed Junco

90
Common Grackle

90
Yellow-headed blackbird

91
Red-winged blackbird

94
Western Tanager

95
Lazuli Bunting

95
Evening Grosbeak

CHAPTER ONE

Bird Life

This chapter reveals the incredible world of birds. The more you know about their world—their struggle for survival, migration patterns, predatory habits—the greater your involvement and enjoyment in bird-watching. Why does a warbler's bill differ so greatly from a macaw's, or the feet of a Harpy Eagle from those of a cormorant? Does plumage change with the seasons? An appreciation of bird anatomy will help you understand why a bird has a particular habitat and prey and has certain behavioral patterns.

Bird anatomy

Birds exhibit a wide range of biological adaptations that distinguish them from other animals; many of these are geared toward flight, the single most distinguishing feature of birds. The anatomy of flying birds is a superb example of the compromise between structural strength and low weight that is needed to achieve efficient flight. Other adaptations help different birds survive in their particular habitat.

In common with other warm-blooded vertebrates (animals with backbones) such as mammals, birds have a rigid, bony skeleton that supports and protects the soft tissues and organs within the rib cage and provides anchorage for the muscles.

The skeleton

Bird skeletons, however, have evolved from the heavy structures of their reptilian ancestors into much lighter, but sturdier, frameworks. The larger bones of flying birds are hollow and reinforced with a network of crossbars—like the truss on a bridge, this structure combines great strength with low weight, giving the bird a low take-off weight and a high power-to-weight ratio, which is vital for efficient flight. Besides being hollow, many bird bones are fused together, which reduces flexibility, but greatly increases strength in order to resist the great forces birds experience during flight, especially during take-off and landing.

Head

A bird's head is small, with lightweight, cavity-filled skull bones, and typically very large round orbits, or eye sockets. Their toothless jaws, which are the foundation of the beak, vary greatly in shape according to the specific, often diet-related, tasks for which these tools are needed (see Birds' Bills, pp. 18–21, and Birds' Feet, pp. 22–23). The underside of the lower jaw consists of a soft throat pouch, or gular region, which is little developed in birds such as sparrows and finches, more obvious in some, such as cormorants, and most evident in the remarkable extensible pouch of the pelicans. The upper bill has openings for the nostrils, usually close to the base. The lower bill allows little or no sideways, or chewing, movement. In many birds, especially swifts, nightjars, and other insect-eaters, the bill is relatively small but the mouth, or gape, is very wide.

Spine, pelvis, and legs

The "wishbone" consists of two clavicles, or collarbones, fused together to act as a strengthening strut that braces the wings apart. Flightless birds and some others, such as the parrots, have a greatly reduced wishbone. A bird's neck varies greatly among species in length and mobility, but in all birds, the bones of the spine are mostly fused together to form a solid, rigid unit that is firmly attached to the large pelvic girdle. The pelvic bones are also fused, and act with the rigid spine to distribute the weight of the body during landing. The lightweight leg bones are operated by powerful muscles on the upper leg, but only by tendons on the lower leg—this sturdy, compact structure acts as a shock absorber during landing. Birds' feet vary greatly in size and form, playing vital roles in locomotion and feeding (see Bills and Feet, pp. 18–23).

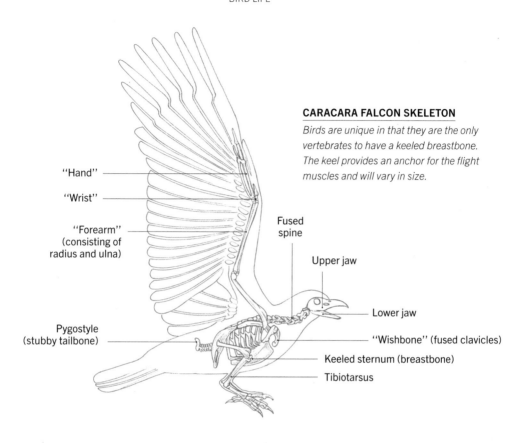

CARACARA FALCON SKELETON

Birds are unique in that they are the only vertebrates to have a keeled breastbone. The keel provides an anchor for the flight muscles and will vary in size.

"Hand"

"Wrist"

"Forearm" (consisting of radius and ulna)

Fused spine

Upper jaw

Lower jaw

Pygostyle (stubby tailbone)

"Wishbone" (fused clavicles)

Keeled sternum (breastbone)

Tibiotarsus

Breast and wings

The breastbone, or sternum, varies from family to family; in gliding birds it is smaller, but in those with powerful, deep wingbeats, such as pigeons, it is large and has a prominent ridge, or keel. The relatively massive breast muscles, or pectorals, attach to the keel, which provides a deep anchorage for the powerful downward sweep of the wings made by contracting the pectorals. The wing is reminiscent of a mammal's arm or forelimb, but with variations: the upper arm is generally embedded in the body or hidden beneath the body feathers. The "inner wing" corresponds to the human forearm (the radius and ulna) and the outer wing, beyond the "wrist," to the "hand," which has several fused digits but a separate thumb (see Feathers, pp. 16–17).

Respiratory system

Birds have developed a highly specialized respiratory and circulatory system that allow them to metabolize oxygen and other substances rapidly so that vast amounts of energy can be generated—relative to their size—to power their wings. Air first passes into the lungs, and then farther into extensions of the lungs called air sacs that reach into the bird's hollow bones, before passing out through the lungs again, allowing the bird to take oxygen on each passage. A bird's heart also beats very rapidly to move the oxygen: a hummingbird has up to 1,000 heartbeats a minute during flight.

Feathers

A feather is an amazing structure. It is made of keratin, a hornlike substance that also forms the basis of our fingernails. It is lightweight, yet strong and durable—essential structural requirements for a bird to achieve flight.

Each feather has a main shaft, or rachis, from which extends a "vane" on each side, consisting of scores of small, interlinked branches called barbs. The base of the shaft is hollow to maximize strength and minimize weight. Together, these make up the basic leaflike shape of the whole feather.

Zipped together

The barbs are held together in a complex series of indentations and hooks called barbules, which interlink in a similar way to Velcro. If the barbules become detached, the feather appears disarranged and ragged. Birds preen regularly every day, each one spending long periods of time gently but firmly sliding its feathers one by one through its bill. This action removes any

debris from the feather, zips the barbs together again, and keeps them in good shape.

Flight feathers are stiff and usually slightly curved. Smaller feathers tend to be wider, and less rigid, especially at the base, where some of the vane is composed of loose and wispy barbs. These barbs at the base of the shaft are almost entirely lacking the barbules that zip the rest of the feather together. They create a soft, down-like layer that insulates against heat and cold.

The main feather tracts

If you spread a bird's wing, you will see that the feathers are arranged in neat zones. Forming the wingtip are large, long, stiff feathers that grow from the fused "fingers" of the "hand"—these are the primary feathers and are of different lengths. Along the trailing edge of the inner half of the wing are the secondary feathers, which are also stiff but shorter than the primaries. These grow from the "forearm." Just beyond the bend of the wing is the bird's "thumb," from which grows a small tuft of feathers that can be raised from the main surface of the wing in flight. This is called the "alula," or bastard wing. The alula is important in helping a bird maneuver in the air at low speeds, and it has long been emulated in aircraft designs.

Each of the main tracts of flight feathers of the wings and tail is covered at its base by a

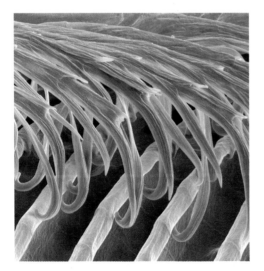

BARBULES
Feathers have a brilliant hook-fastener construction, with thousands of tiny hooks clinging to minute stays.

WING ANATOMY

The main tracts of flight feathers on a bird's wings, such as the primaries and secondaries, are overlapped by the coverts—various series of smaller feathers that merge with the main tracts to form an aerodynamic surface.

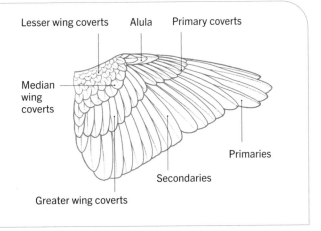

Lesser wing coverts   Alula   Primary coverts

Median wing coverts

Primaries

Secondaries

Greater wing coverts

row of smaller feathers—the coverts, which cover the gaps at the base of the flight feathers, creating a smooth, solid surface for efficient flight. The contour feathers on the bird's body are less obviously, yet still precisely, arranged, cloaking the body to give it its streamlined shape. There are areas of a bird's body that are less defined by feather tracts than by color and pattern, such as the forehead, crown, nape, cheeks, breast, and flank.

Wear and tear

Feathers gradually become ragged and lose color. The pigments that provide most of the colors in plumage (see Colors and Patterns, pp. 24–25) fade with age, especially if they are exposed to strong sun or saltwater. These exposed parts become paler and duller while the base, usually covered by other feathers, remains bright and fresh.

Molt

Feathers, once grown, are dead structures that gradually wear and deteriorate. To counteract this, birds molt their feathers regularly, shedding and replacing them symmetrically, and usually a few at a time, to maintain maximum efficiency. Some ducks and geese,

however, drop all their flight feathers at once and are unable to fly until new ones grow. Most birds, though, shed their feathers one by one in a regular, fixed sequence. Watch flying birds and look for gaps in the shape of the wing, or for new, fresh, dark feathers growing among paler, rough-edged, old ones.

Replacing feathers consumes a lot of energy, and molt must therefore occur when energy-rich foods are readily available and when the bird is not involved in other energy-sapping activities. It is no accident that the complete annual molt of birds occurs after breeding and usually before migration.

Not only does molting refresh a bird's plumage, it can also change a bird's appearance according to age and season. Dull-brown young birds, for example, become brighter as they gain adult feathers. Drab winter plumage can be changed for brighter breeding plumages. It is not always so simple: ducks, for example, pair up in winter and look brightest in the coldest months. In summer, the male loses his ability to fly and molts into a faded, dark pattern—called "eclipse" plumage—so that he is more camouflaged and less likely to fall prey to a passing hawk or fox.

Birds' bills

The bills of birds occur in a tremendous range of shapes, reflecting the adaptation of each species to its unique mode of obtaining food. Many birds also have specially shaped or colored bills that they use in courtship rituals.

COMMON RAVEN
The raven uses its heavy, arched, powerful bill for all-round foraging and tearing up tough items before eating.

ICTERINE WARBLER
Shallow but broad, this bird's bill is shaped for seizing insects.

CHANNEL-BILLED TOUCAN
This bird has a remarkable bill, which is very large but extremely lightweight, for reaching fruits on thin twigs.

A bird's bill consists of an upper jaw, which is fixed on the skull and almost immobile, and a lower jaw that is articulated like ours and opens to reveal a wide gape.

Hornlike sheaths
Each jaw has a bony base. The lower base is formed by two bones, which are fused together where they meet near the tip of the bill. The space between them is filled by the soft throat and chin (or sometimes a fleshy pouch). These bones are covered by hard, hornlike—or sometimes leathery—sheaths that give the bill its detailed shape.

On a few birds, such as the puffin, the sheaths are shed or modified after the breeding season, allowing for marked differences in the size of the bill, according to the season. In most birds, seasonal changes are limited to alterations in color. On birds such as herons, these color changes can be brief but rapid, with a visible increase in the intensity of color during periods of excitement in spring courtship.

Flexible bills
While the upper jaw is fixed, the tip of the long upper jaw extending from the skull can be surprisingly flexible. Even birds such as cormorants can raise the upper bill to an unexpected degree in a wide "yawn." Snipe and woodcock have especially sensitive bills, and they can detect prey while probing deep in soft mud or soil. Once a worm has been found, by touch, the bill is flexible enough to allow the tip

to open and grasp the prey, which can then be swallowed without a pause in the deep probing.

Seed-eaters' bills

Sparrows and finches have broad, deep, triangular bills for feeding on hard seeds. They vary, nevertheless. A goldfinch has a fine, pointed, triangle-shaped bill for probing into complex flower heads such as thistles for seeds. The bill of a crossbill is literally crossed at the tip, which allows it to reach inside of the scales of pine and larch cones to extract the seeds within with its tongue. Most finches have "middle-sized" triangular bills for crushing tough seeds, and their bills have sharp cutting edges to de-husk them. At a feeder you can often watch a finch manipulating a seed with its bill and tongue, peeling off the outer husk and nibbling at the kernel within.

Insect-eaters' fine tools

Insect-eating birds such as warblers have fine, narrow bills for picking tiny prey from twigs and foliage. Flycatchers have broader bills, often fringed with stiff bristles, and wide mouths so they can catch insects in flight. Swifts and nightjars have minute bills, but huge mouths, which open wide to catch insects in the air. But many birds have "all rounder" bills, and even sparrows can catch insects when they wish to, and scoop up vast numbers of aphids when feeding their young.

Starlings have stout, strong, pointed bills with especially adapted muscles that they use to probe into grass and open their bills to get at the larvae of chafers and other prey. A thrush has a stronger bill than a warbler, and it is longer and more slender than a sparrow's: they eat worms, grubs, seeds, and all kinds of fruit. Its bill is more of a tool for all trades than a specialized instrument.

GREEN WOODPECKER
The dagger-shaped, chisel-like bill of this bird is used for digging into ant-hills and chipping out nesting cavities in tree branches.

NORTHERN CARDINAL
This bird has a thick, triangular seed-eater's bill.

BLUE-AND-YELLOW MACAW
The deep, arched, hook-tipped bill of this bird is used to rip open fruits and seed pods.

GOLDEN EAGLE

Typical for a bird of prey, this eagle has a hooked bill, which is used for tearing apart meaty foods.

GIANT HUMMINGBIRD

This bill is specialized for taking nectar from flowers and snatching tiny insects.

COMMON REDSHANK

The slim, sensitive bill of this bird can probe into wet mud and sand.

Hooked and toothed bills

Birds of prey have a hook-tipped bill, also known as a beak, for tearing prey apart. Few of these birds capture prey in their beak—they normally use their feet to snatch their victims— but many kill prey with a sharp bite. A falcon's beak has a small "tooth" on the sheath of the upper bill to help sever the neck of a mouse or small bird. Vultures have bills that are designed for the food in their local area: the big Black and Griffon Vultures (true members of the Falconiformes) of Europe have heavy bills that can tear into the hide of dead animals, while the Turkey Vulture of the Americas (related to storks) and the Egyptian Vulture of southern Europe (another Falconiform bird) have narrow, fine bills for probing deep inside carcasses made of weaker tissue.

Other birds also have hooked bills. Parrots are primarily fruit and seed eaters, and their thick, hooked bills allow them to cope with large fruit and seed pods. Mergansers not only have hooks, but also tooth-like edges to the bill, to keep a firm grip on slippery, muscular fish. Gannets, kingfishers, herons, and egrets manage to eat their fish diet quite well without either feature, grasping prey in their powerful, pointed, sharp-edged bills before turning them to swallow the fish head-first so there is no risk of choking on extended spines or fins.

Probes and hammers

Shorebird bills are greatly adapted to habitat and food preferences. Curlews have down-curved bills, perhaps so they can easily see what they are doing with the bill's fine tip. But they also probe and twist the bill, and the curve might allow them a better chance to detect prey. Godwits, snipe, and dowitchers, however, probe adequately with long, straight bills.

Plovers take food from the shore, often on mudflats, using a shorter, thicker bill; a Dunlin picks from, or just beneath, the surface of mud and sand with a longer, fine-tipped bill. A turnstone has a strong, slightly up-curved bill that it uses to move pebbles and seaweed in search of hidden invertebrates. An avocet's up-curved bill is swept sideways through shallow water to capture prey near the surface, while a spoonbill's broad bill with its wide, flattened tip is swept—slightly open— from side to side through the water until a fish is touched and the bill grabs it as with a pair of salad tongs.

Oystercatchers can have a pointed bill, used to slip inside shellfish to cut the muscle that holds the shells tight together, or a blunt-tipped bill, which is used to hammer the shells to pieces.

RED-BREASTED MERGANSER

With serrated edges for grasping fish, this is a typical sawbill duck's bill.

MALLARD

This bird uses its broad bill for "dabbling" in shallow water to pick up seeds and aquatic creatures.

AMERICAN DARTER

This fish-catcher bill can grab or stab fish.

Birds' feet

Birds rely not only on their wings to get around—they use their feet for walking. Their feet are adapted for walking styles as well as for gripping tasks, whether it be an eagle snatching a fish or a robin perching in a tree.

COMMON PHEASANT
The strong, stubby feet of this bird are adapted for walking.

CARRION CROW
This bird has strong feet adapted for perching and walking.

Most birds have four toes, three pointing forward and one backward, to allow for a good grip on a perch, such as a twig or branch. Small songbirds that walk or run on the ground, such as larks and pipits, have a very long hind claw. Some larger birds that walk or run in open places have only three toes, with the hind toe lost or reduced to a stub.

The hind toe is also much reduced in ducks, geese, swans, and gulls; instead, the front three toes are joined by leathery webs that give a stronger push against water for more powerful swimming. Gannets, pelicans, cormorants, and their relatives reveal their close relationship by their feet, on which broad webs join all four toes.

Falcons have strong toes and sharp, arched claws, with which they grasp their prey. Bird-eating harriers and hawks have longer legs, much longer toes, and needle-sharp, curved claws that enable them to capture birds in flight and kill them by puncturing vital organs.

NORTHERN JACANA
The exceptionally long toes of the jacana help to spread its weight on floating leaves.

ROCK PTARMIGAN
Feathered toes give this bird better insulation in snow.

GREEN WOODPECKER
The outer toe of this species splays outward and backward to give good grip on a rounded branch.

HOUSE SPARROW

Typical for perching birds, the sparrow has three toes pointing forward and one back.

AMERICAN HARPY EAGLE

Its huge feet, strong toes, and arched, sharp claws enable this bird to kill large prey such as monkeys.

Waders

Shorebirds may have very long legs for wading in shallow water. Plovers have just three toes, while sandpipers have four, longer, more slender toes. Avocet toes are joined by partial webs. Similarly, rails and crakes may have very long, slim toes to take their weight on soft mud and floating vegetation—yet the closely related coots have broad lobes along each side of each toe, as do the grebes. These fold flat when the foot is pushed forward through water, reducing drag, but open out as the foot is pushed back, to give forward thrust.

GREAT CORMORANT

All four toes of this bird are joined by webs to provide powerful underwater propulsion.

BLACK COOT

The lobes on this bird's toes spread out on back stroke and fold away on forward stroke for efficient swimming.

OSTRICH

Its feet are designed for long-distance walking and running at great speed.

GREY HERON

Its long legs and toes allow this heron to wade in the shallows and stand on soft mud without sinking.

Colors and patterns

Feathers occur in just about any color variation you can think of. Their patterns and colors can be amazingly complex. Yet many birds have a basic similarity, especially within families.

Remarkable consistency

If you look at 10,000 Black-headed Gulls at a reservoir roost, there will be almost no variation among them in the color of their upperparts: all are the same pale, silvery-gray. Watch the thrushes in your garden and, year after year, they will each have exactly the same shade of brown on the back. There are individual differences, but the essential truth is that birds of one species are usually amazingly alike. This helps bird-watchers, because it makes the birds easier to identify. Variation within a species is often linked to geography. When one species has a wide geographic range, those at one end can be a little—or, rarely, a lot—different from those at the other. In many geographically variable species, there is a smooth gradient of variation (such as in size or color) from one extreme of the range to another. Such a smooth gradient is called a "cline."

Differences according to age

While most young birds look a little unlike their parents, sometimes the differences can be striking. Young gulls are mostly drab, mottled brown, and quite unlike their sharp parents. They need to stay hidden, and their brown color provides good camouflage—an adult will treat another individual in adult plumage as a potential competitor and will react aggressively toward it. It will react differently to a bird in juvenile plumage, however, recognizing that it is only a juvenile and poses no threat.

Differences according to sex

Many species, including wrens, Blue Jays, Red-headed Woodpeckers, and some thrushes, have males and females that are identical. Others have marked sexual differences: Northern Cardinal, Eastern Towhee, most woodpeckers, ducks and gamebirds, such as pheasants, are good examples. In some polygamous species, such

NORTHERN SHOVELER

There is no discernible difference between the shades of rusty-red, dark green, brown, and blue on shovelers found across the Northern Hemisphere.

as Ruffs and Ruffed or Sage Grouse, males display together in spring so that females can choose the best mates. In Ruffs, the difference in color is heightened in spring, but in other species, such as pheasants and peacocks, the difference remains all year round.

Differences according to season
Spring and summer see many male birds at their finest: they are in breeding or courtship plumage. During this time of year they need to impress females and also ward off rival males without having to fight—they can do so by looking brighter, fitter, and more dominant than their rivals.

A Black-headed Gull gains its hood in late winter; the dark hood develops from the back of the head forward so that it is complete, and most intimidating to rivals, by early spring. The brilliance of wildfowl males, such as goldeneyes, Mallards, and mergansers, is best seen in the winter, when the sexes pair up, often before they migrate north to breeding areas. Yet geese, which are closely related to wildfowl, have no obvious differences between male and female during the summer or winter. There is clearly no "right" way to use color and pattern, and there are many different strategies that work for different species.

Camouflage
Birds do not have the ability to camouflage themselves as well as insects that look like leaves or fish that look like seaweed. They must be able to fly and live active lives, so they rely on their colors and patterns, rather than outlines, to help them hide. But some birds have developed wonderful camouflage: woodcocks blend in with dead leaves on the forest floor, and grouse blend into their heathery backgrounds exceedingly well. Many

KINGFISHER
Many colorful birds rely on the structure of their feathers to separate white light and produce brilliant hues, rather like the pits on a compact disc.

birds the world over have darker feathers on the back than on the belly. The natural tendency for a bird's upperside to be lit by stronger light and the underside to be in shade helps "flatten out" the bird's colors and makes it more difficult to see it. Look at a thrush or robin and you will see that the flank—just where the sun tends to catch the fluffed out feathers—tends to be a little darker than the belly. This pattern of dark above and light below is known as countershading.

Showing off
Swans are so big and so strong that they really have no need for camouflage. Instead, they are dazzlingly white and brilliantly obvious to other swans from a mile away. This simple statement communicates their presence to all other birds.

White plumage is used in other ways, too. Gannets plunge from a height to catch fish. If one dives, the rest notice the bright white bird creating a big white splash, and they hone in on the spot, congregating rapidly above a shoal of fish. Colors and patterns are used to display aggression, too. A robin spreads its red breast feathers to antagonize other robins. Usually such a show of strength and fitness is enough to avoid a fight: the bright color helps reduce the need for more drastic action.

Food and feeding

The great diversity of birds is largely due to the many different kinds of foods that they can exploit, along with the variety of their habitats and the isolation of species by geographic barriers. Most species of birds are adapted to feed on something very slightly different from their close neighbors, so that competition for prey is kept low.

In general, birds eat feverishly early in the morning after having fasted all night. Then they rest, feed casually during the day, then feed feverishly again before going to roost for the night.

Food from the trees

Brown Creepers feed on minute insects and spiders, including their eggs and larvae, on the bark of trees. They explore the tiniest crevices using their thin, curved bills. Nuthatches also explore the bark but are more likely to take larger food and eat far more seeds and berries. They have stronger feet for a better grip, which makes them more agile, and bigger bills to tackle larger food items—they can even hammer open tough nuts.

COMMON SNIPE

The bill is flexible but strong enough to open underground and grasp a worm.

Chickadees and titmice also explore tree bark and foliage. The Tufted Titmouse tends to prefer the bark of the trunks and bigger branches of trees, as well as finding food on the ground below, while smaller chickadees feed on slimmer branches and twigs, looking for smaller prey. In this way, these different species can move together as a flock, keeping their eyes open for possible danger, while keeping out of each other's way and eating different food.

Chickadees find an abundance of food at different times of the year and hide a lot of it by pushing it into places that they are likely to search again later, such as bunches of pine needles. These stores of food can then be saved for the winter. While chickadees find stored food again by chance, various crows, jays, and magpies locate their stored food by design. They bury nuts, scraps of meat, and acorns and can remember where they put their winter stores months later.

Beneath the surface

Shorebirds feed in rhythm with the tides, so their feeding times fluctuate from day to day, always coming at low tide. Wading birds on a beach use different feeding strategies too. A Semipalmated (with toes webbed for part of their length) Plover picks small shellfish and crustaceans from the surface of the mud, while a Semipalmated Sandpiper or a Dunlin picks from an abundance of minute snails

FLYCATCHER

While warblers mostly have fine, thin bills, flycatchers have broader bills with bristles around them, making it easier to snap-up flying insects.

barely hidden in the wet mud at low tide. A Red Knot has a longer bill and probes a little more deeply. Godwits probe deeper still for small worms, while curlews, which have very long bills, probe deep down for bigger lugworms, as well as taking ragworms, shellfish, and small crabs from rock pools as the tide falls. Each in its own way is adapted to feed on different food, or in different situations, so that all can live together on the same beach.

Climate change

Many seed-eating birds, such as finches and sparrows, need to eat insects during the summer. The adult birds manage to survive on only seeds, but their fast-growing chicks must have energy-rich, high-protein food, and the best source is insect food. Finches feed thousands of leaf-eating caterpillars to their young in the nest. Recently, tits have had poorer breeding seasons, and this may be because they are nesting at the "usual" time, while their caterpillar food is emerging two or three weeks earlier due to the effects of climate change. Global warming is wreaking havoc in some places as the reproductive cycles of birds and their prey are becoming out of sync.

NO CONTEST

A remarkable means of avoiding competition is seen in birds of prey, especially in bird-eating hawks, such as Cooper's and Sharp-shinned Hawks—the female is up to one-third bigger than the male. This means that the sexes eat prey of different sizes and so do not compete with one another, allowing them to survive together in a smaller area without straining the supply of food.

How to feed birds in the garden

Feeding garden birds has become a multi-million dollar business. The trend began in North America and followed later on in Europe. Today, the new foods that are being used in bird feeders are attracting different kinds of birds, and present the food in ever more sophisticated ways.

Bird lovers feed garden birds for two main reasons: to help birds survive in a tougher environment, and to see them up close. Fortunately, the two are mutually beneficial.

Positioning a feeder

To get a good view of birds feeding in your garden, the feeder must be placed so that you can see it from a normal position within the house. A feeding tray must be high enough that you can easily see it while sitting down in a favorite comfortable chair. It is no use if it is hidden below the window level.

The feeder should be placed away from any cold, windy channel between walls or buildings—a little shelter is always welcome. It is also a good idea to position it away from a footpath, where the birds may be disturbed.

Any feeder is vulnerable to predation: if you put out foods to attract birds, you will attract hawks and other predators, too. A bird feeder covered in tits provides an easy meal for a hawk. There is no way to eliminate this danger, but you can do things to minimize the risk. The most effective solution is a dome of large-mesh chicken wire placed over the feeder, which will allow smaller birds to enter while keeping predators out. Netting is likely to be chewed through by squirrels. Suspending unwanted CDs from loose strings on the feeder may distract both hawks and the songbirds you want to attract—but this deterrent will not last.

If you place a feeder close to a bush or tree, it gives small birds a sporting chance of diving to safety. On the other hand, positioning a feeder by thick shrubbery or a flowerbed might invite unwanted attention from the local cat, the garden bird's greatest enemy.

When to feed birds

You can feed birds all year round. Birds need food in winter to help them survive the cold nights, but spring can also be a difficult time for them. Those that rely on seeds and fruit can find that natural supplies are at their lowest ebb by late spring and a free handout of birdseeds and peanuts can be a real boost just before the breeding season.

DIFFERENT TECHNIQUES

Try putting fat and cheese in crevices in tree bark, and scatter grated cheese under bushes for shy birds that don't come to tables. Feeding birds in several places will attract a greater number and variety, and allow birds to feed without constantly squabbling.

KEEP IT NATURAL

It is vital to include natural bird foods in your garden and to have a variety of plants that can provide them. House sparrows are declining fast in many parts of the world. In winter, they have fewer seeds to eat as herbicides have reduced the number of weeds and high-tech harvesting has minimized the spillage of grain. In summer, they are unable to rear young, because pesticides and low-maintenance gardens with fewer shrubs mean a lack of insects.

FEEDER TYPES

Rigid spiral metal feeders (above left) are safe, but springy ones may trap birds' feet. Solid-mesh feeders (above center) are ideal for bulky foods; special tubes (above right) are needed for finer seeds.

Remember that young birds are being fed in the nest during spring and summer. Although most birds will not feed their chicks food that might endanger them, if natural food is scarce, they may resort to feeding them foods such as peanuts. Baby birds can choke on large peanuts, so make sure they are crushed and crumbled, or wrap them in a strong mesh that is fine enough to make birds peck pieces from them, rather than take them away whole.

Types of feeders

Do not use flexible feeders. One neat design is the coil spring feeder with a base and lid—but make sure the coil is very rigid. A flexible coil may catch the legs and feet of feeding birds and injure or kill them. Peanuts have long been sold in plastic mesh bags, which can be hung outside as ready-made feeders. These are generally fine to use, but increasingly people have become aware that they can trap and kill small birds. Birds can get their feet and even their tongue tangled in the mesh so that they hang from it until they die. It is best to avoid such bags and instead place the nuts inside a rigid metal mesh basket or a plastic tube with special feeding ports.

Foods to offer birds

Feeding birds in a garden can be great fun: you can experiment with different foods and the way you offer them. There are types of bird foods available to suit every budget, from expensive seeds to cheap kitchen scraps. Just remember to keep everything clean. Piles of old bird droppings or decaying food can promote disease and wipe out the very birds you are trying to save.

Plant shrubs and herbs—especially species native to your area—in the garden to help birds find food. Many of these plants provide nectar and attract insects, which birds eat. Others produce fruits—all kinds of berry-bearing shrubs are excellent for birds, especially cotoneasters, berberis, and hawthorns, as well as pyracantha and holly. Such natural foods are good for birds and keep a natural balance in the garden.

If you have feeders and a birdtable, or a ground feeder, you can do a great deal to help birds. They need energy, which often means fat: full-fat cheese, cooked bacon rind, suet, and animal fats make perfect foods and can also form the basis of "bird cakes."

To make a bird cake, use the fatty material to bind together nuts, seeds, fruits, and scraps. Squeeze or pour the mixture into containers such as yogurt pots and coconut shells. Hung in trees or from the birdtable, these are great feeders that offer food for winter birds. Do not put such fat-based foods out in the summer heat. The fat melts and can cause feather damage and loss as the oils soak into the bird's plumage.

The right foods

Bread is a good type of food to provide birds in small quantities. Brown, damp bread is preferable—dry, crusty bread is often neglected. Stale cake crumbled on the birdtable or on the ground is often a better choice. Kitchen scraps of all kinds, from uncooked or cooked pastry to bits of fruit, will usually be accepted.

Apples also make good food, especially if they are cut into small pieces or halved and scattered on the ground for thrushes, or put on the birdtable. Scatter them in several places if you have space, so that a number of birds can feed without fighting.

Peanuts are a popular, traditional food for birds and remain ideal, especially for tits when hung in a mesh or special tubular feeder—but many finches and other birds will eat them, too. Woodpeckers often find them and come frequently to feed on them.

Sunflower seeds are excellent, oily food for birds such as the larger finches, and are a great alternative to peanuts. Smaller finches will also eat oats and millet scattered on the ground or a table or placed in a feeder. Nyjer, or niger, seed is a much finer seed that requires special feeders. It may not work, but when it does, it can attract goldfinches and siskins and keep dozens of them coming back for weeks. It is, however, quite expensive.

Wild bird seeds come in many mixtures, from cheap seeds to pricey high-protein mixes. Cheaper seeds have a lot of large grains, which are not eaten by many birds other than pigeons, and may even be padded out for bulk with substances such as broken dog

FEEDING TIPS

When you feed the birds, take the following steps to provide a safe and healthy feeding environment.

- Position feeding stations in different areas of your yard to spread birds around and avoid competition, stress, and disease.

- Clean your feeders regularly with hot water, and let them dry completely. Keep areas under and around the feeders clean.

- Keep seed clean and dry and watch that it doesn't get moldy. If there is a lot of waste, reduce the amount of food you put out.

- Use a seed blend that is designed for the feeder you have and the type of birds that come to that feeder.

- Offer seeds in a feeder rather than scattering seed on the ground.

- If possible, move your feeding stations periodically, so there will be less concentration of bird droppings.

- Always wash your hands after filling or cleaning your feeders.

- Place bird feeders in locations that do not provide hiding places where cats and other predators can wait to ambush the feeder. Bird feeders should be placed 5–12 inches (12.5–30 cm) from low shrubs or bushes that provide cover.

- Place the feeder 5–12 feet (1.5–3.5 m) from a brush pile or bushes to provide a place for birds to take cover in the event of danger.

biscuit. Try to choose better quality foods from reputable sources.

Don't feed birds salted foods, such as salted peanuts, potato chips, or salty bacon. Also avoid giving desiccated coconut and other dried foods that may swell after they are swallowed.

Water

Fresh water is vital for birds, all year round: even in the depths of winter they need to drink and they need to bathe. Put out a dish or fill a birdbath each day. Keep the water clean, and never use any artificial substances to prevent the water from freezing.

Hygiene

Cleanliness is important for birds and every bit as vital for you. Use rubber gloves when handling feeders and cleaning tables; use a stiff brush and keep it solely for the purpose of cleaning your birdtable. Move your feeders around every so often to avoid a buildup of droppings and waste. Now and then, wash them in a weak solution of ordinary bleach and rinse them clean.

Nests

The main purpose of any bird is to reproduce and ensure the survival of its species: to do so, it must find a mate, and females must nest, lay eggs, and see that their chicks are reared. A nest is simply a receptacle—or even a mere scrape in the ground—in which the eggs are laid and incubated until they hatch.

In many species the young remain in the nest until they can fly. Once the nest has fulfilled its purpose—by which time it may be an unsavory and unhygienic place containing various parasites, droppings, and uneaten scraps of food—it is usually neglected. Some larger species, however, build more substantial nests, which are refurbished year after year—the nests built by eagles and Ospreys may become very large.

EARTH NEST

Terns, gulls, and wading birds lay their eggs in a scraped hollow in earth or sand.

EAGLE NEST

Many big birds of prey reuse big stick nests, which can become enormous over many years.

Nest uses

A nest is the place where a bird lays its eggs and incubates them until they hatch. It is not a "home" used by birds at other times, although some hole nests are used as roosting sites in winter. A nest or nest site may be used many times over, or a new one may be made for each set of eggs. In general, small birds use new nests for each clutch during a season, as the nest becomes a dirty and unhygienic place after rearing a family: it tends to harbor many parasites, such as mites, ticks, and fleas, probably has waste food in and around it, and is soiled with droppings. Larger birds, which have just one clutch each year, are more likely to reuse a nest in successive seasons, provided at least one of the pair survives: in some species, such as Ospreys and Golden Eagles, nests may be used by generations of birds over decades, and others, such as Peregrine Falcons and ravens, use the same piece of cliff, if not the same ledge, for many years.

Birds with no nest

Not all species build a structure for the nest: some simply lay eggs directly onto the bare ground, onto a cliff ledge, or inside a hole, without any nest material being added. Kingfishers lay onto a bare floor inside a deep tunnel, but as the eggs hatch and the young grow, their nest chamber fills with bits of fish and undigested fish bones, as well as foul, semi-liquid droppings.

Some birds such as guillemots and other seabirds, and Peregrine Falcons, lay eggs on a bare ledge, or at best a scrape in earth or gravel naturally collected on a cliff ledge. Hygiene here is also suspect: seabirds in colonies may be splashed with droppings from birds on ledges higher up the cliff. Birds of prey, however, while they may build up unwanted food that simply rots around the nest, usually at least keep the

WOODPECKER NEST

Even firm, healthy wood can be chipped away by a woodpecker as it excavates its round, deep nesting hole.

LEDGE NEST

Some birds of prey, including Peregrine Falcons, nest on earthy ledges on the face of cliffs.

nest free from droppings, and the young birds quickly learn to back up to the side of the nest and void their excreta over the edge. Other birds, such as plovers, add a few scraps of vegetation, shells, or stones to a scrape on the ground or in a sandy beach and the addition of such objects can become ritualized.

Nest materials

While falcons make no nest, hawks do so, usually in trees. Eagles and Ospreys can build up huge nests using sticks, which accumulate over decades of repeated use, creating incredible structures on cliffs, poles, or in treetops. They also bring green foliage to the nest all summer, perhaps to help keep down flies and pests in a nest full of rotting food.

In those birds that do build nests, often both sexes share the work, but the female may add the final lining. In many cases the male makes the initial foundations. Male wrens make several nests as part of their courtship routine:

SONGBIRD NEST

A typical small bird's nest is a cup-shaped structure, often with a softer lining.

they attract females and show them a choice of nests, perching nearby and waving their wings as they encourage the female to inspect each site, but the female makes the final choice of which one to use.

Many nests are simple structures of sticks and twigs, usually with a lining of smaller materials and perhaps feathers, hair, or fine plant fibers. Such a nest made by a small songbird can be completed within two days: other more complex structures may take a week or two to complete. Magpies add a protective dome to their "fortress" nests. Small songbirds make more complex structures, usually cup-shaped, with a softer lining; those of the thrushes, such as American Robins and the European Blackbird, have a strong, hard mud layer inside a basic structure of grasses and twigs. Some, such as the Long-tailed Tit's nest, or the smaller, hammock-like nests of Goldcrests and kinglets, are beautiful, delicate, and made of lichens and moss held together with spider's webs. These nests stretch as the brood of chicks grows. Wildfowl pluck down from their own bodies to add a warm, soft layer to the inside of the nest.

Hole nests

Many birds nest in holes of some kind: woodpeckers chip out a hole in a tree, while kingfishers dig into an earth bank. Birds as varied as chickadees and tits, owls, and kestrels occupy a hole in a tree—either a natural hole or one made by a woodpecker. Nuthatches of several species plaster up the entrance hole with mud, until it is just big enough to allow them in, but excludes larger predators. They will also plaster mud between a nest box and the bole of a tree, which perhaps insulates the box and provides greater stability. The plastering habit is probably simply instinctive and used whether it has a necessary function or not: nuthatches cannot resist doing it.

Artificial nests

Birds that nest in holes can be helped, even in gardens, by the use of artificial nest boxes. In the U.K., Blue and Great Tits are common nest box occupants, while populations of Pied Flycatchers have been increased in many woodlands by the provision of boxes where natural holes are few. In the U.S.A., Purple Martins will use large "apartment house" boxes on special poles, and bluebirds can be helped by the provision of boxes in and around gardens. Special boxes can be used by owls and kestrels, while artificial structures that provide shelter for chicks have helped rare birds such as Roseate Terns improve their breeding success on nature reserves, where severe weather and predators otherwise take their toll.

Nest boxes

Nest boxes can easily be made to standard sizes and designs from wood, or "woodcrete"—a mixture of sawdust and cement—which helps to protect eggs and chicks from the attentions of woodpeckers and other predators that can gain access to a normal wooden box. While metal shields around the entrance may deter some predators, woodpeckers can still dig their way in from the side unless a more resilient material is used.

NEST BOX

These structures can be made in many shapes and sizes according to the bird you have in mind: this one would suit a small hole-nester.

TENEMENT BLOCKS

The Purple Martin has been exclusively nesting in this style of nest box for nearly a century in the eastern part of North America.

Song, courtship, and display

During the spring, a territorial bird (usually a male, but sometimes a female or a pair) has to find a territory that will support himself, a mate, and their young. Birds use song to attract a mate, repel rivals, or both. Courtship involves ritualized displays that help break down a bird's instinct to maintain individual distance and strengthen a pair bond. Such displays often have different meanings from species to species.

LAPWING

This shorebird combines striking colors and shapes, distinctive calls, and a wild, tumbling flight in its displays.

By singing, a male bird tells other males that an area is occupied—it is a clear message that provides birds with space that is free of competition and interference—and the space usually comes as a result of song rather than potentially injurious physical conflict.

Song

Songs are specific to each species. Basic songs, such as a wren's, are simple and have little variation. Longer, more sophisticated songs develop with age as the bird adds to its repertoire, to show how mature and experienced it is. A rich variety of songs is used by the Marsh Warbler in the U.K. and the Northern Mockingbird in North America, which add to their repertoire by imitating a vast range of other birds—maybe 100 or more. The Marsh Warbler even includes mimicry of birds it encounters in its winter in Africa. The reason for this behavior is unknown—in males, it may be simply a method of displaying virtuosity to the females. At the other end of the scale, a young bird deprived of hearing others of its species will grow up using an ordinary, simple, "low-quality" version of its species song, reflecting the song's inherited nature but also the influence of competition with and imitation of others. If two birds that have not yet claimed a territory both want to settle in the same area, and both have a more or less equal interest in and access to it, they may fight ferociously.

Male birds generally fight other males, and females fight other females—sometimes, but rarely, to the death. If two males are both settled in their own territories and meet at the boundary between them, they will call, sing, and display to avoid fighting. If one bird trespasses, the other, defending, bird will usually manage to drive it away.

Courtship

In many songbirds, and some other birds such as terns, the male feeds the female during courtship and early in the breeding season. Such courtship feeding aids in pair bonding, but also helps the female survive a period of stress, when she is using huge amounts of energy to lay a large clutch of eggs.

Display

In large birds of prey, such as larger hawks and eagles, aerial displays take the place of song. Sounds do not carry far in the huge territories of these birds, but they can see each other from great distances. These birds spend hours soaring over their nesting areas so that others of their species can see, and keep away.

Not all birds form long-lasting pair bonds when breeding. Game birds such as grouse and pheasants are polygynous, meaning that one male mates with two or more females; the males fight and show off but it is the females that select a mate. Such fights and displays are highly ritualized, and often take place at traditional sites that have been used for decades. Black Grouse in Europe and Greater Prairie

Chickens in North America, for example, collect in spring at a place called a "lek." The males strut, spread their tails, jump in the air, make special sounds, and occasionally fight. The females look on, deciding which is the "best" male to father their young.

WILD TURKEY
A big, healthy turkey is an impressive sight even to us. Its performance must impress both rivals and potential mates.

Calls and postures

Bird calls have different functions from song. Song is used, primarily by male birds, to warn other males to keep away or to attract females. Females of some species, such as the Northern Cardinal in North America, sometimes sing to defend a territory against other females, but their song is more muted than that of males. Calls, on the other hand, are used by birds of all ages and both sexes.

Some calls are simply used to keep in touch, while others give particular messages, even to birds of other species. Certain postures and actions function in a similar way.

Contact calls

A flock of small birds moving through the woods during the winter keeps more or less together in order to maintain vigilance against predators and to improve the chances that one member of the flock will find food that all can exploit. Birds within the flock keep in touch by rather quiet, simple, short call notes, called contact calls.

Such calls often have a long vowel sound but, to our ears, no obvious consonants—they do not have "hard" endings or "shapes" to the sound, and they can often be written down as

"eeee" or "sseee." Similar notes, but with a greater intensity and a sharp or metallic quality, give warning of a predator. These calls are often above the frequency range that is easily heard by a large hawk or falcon, but well within the hearing of small songbirds.

Identifying calls

The lack of a hard edge to call notes also makes them exceedingly difficult to locate. Bird-watchers often hear calls but find it hard to locate the bird that is making the sounds. This allows a small bird to warn others that a predator is near without putting itself at risk. House sparrows use different calls to indicate aerial threats and ground-based threats, such as a cat. If you become familiar with the alarm notes of different birds, you may find that you can recognize the call a bird makes to identify the presence of a hawk, and may improve your chances of seeing these birds of prey.

MOORHENS

The splash of white beneath the tail is used as a means of communication between mates and also between rivals.

ACTING UP

A "broken wing" display is designed to lure predators in pursuit of easy prey away from eggs or chicks.

Warning calls

Adult birds use particular warning calls for their chicks, too. "Keep still" is the obvious meaning of a call when danger approaches. Young birds make long, loud, whining, or squealing notes when they are hungry. One theory to explain such behavior is that it "blackmails" the parents into feeding their chicks to keep them quiet, otherwise, the sounds would attract predators and weeks of effort spent rearing the young would be wasted. An extreme example of this is the young cuckoo, which is reared by foster parents and has a kind of "super" food-begging call. It is so effective that even passing birds that have nothing to do with the cuckoo will stop to feed it. Its loud, wheezing, begging note seems to be an irresistible stimulus.

Small birds make loud alarm calls when they discover a predator, such as an owl roosting in a tree by day. They risk great danger in drawing attention to the owl by "mobbing" it, bringing in a crowd of birds of several species that join in the hue and cry. This may be to make sure that all birds in the area know that the owl is there, or it may help to drive the owl away, or it may even teach young birds that owls are dangerous. Another theory is that such a noise attracts even bigger creatures—such as people—that scare the owl away when they come to investigate.

Postures

A good example of a posture is the injury-feigning of birds such as plovers and skuas. If a predator such as a cat or fox threatens the nest or young, the parent will flutter along the ground, luring the predator away by pretending to be injured and unable to fly. The potential predator misses a meal when the adult judges its young to be safe, suddenly "recovers" and flies away.

Eggs

Birds lay eggs and incubate them externally, in nearly all cases by applying heat from their bodies, until the young bird breaks free. Some small songbird eggs hatch in fewer than 10 days, but the eggs of some species, particularly many seabirds, may require 50 days or more. Albatross eggs, for example, take as long as 80 days to hatch.

Bird eggs contain a developing embryo and various components that nourish and protect it—the yolk, egg white (or albumen), and an air cell. These elements are encased in a more or less smooth, thin, rigid calcium shell of surprising strength that keeps out water, but allows in air. Eggs may be almost round, symmetrically oval or, most often, broader at one end than the other. The narrow end of an egg may be blunt or rather pointed, as in the case of some ground-nesting birds like the Killdeer in North America, which lay four eggs and arrange them neatly, with the pointed ends directed inwards.

Egg design

As the egg is formed, colors and patterns that are characteristic of the species may be laid down on the shell. These sometimes help to camouflage the egg when it is left uncovered in the nest. Hole-nesting birds, including owls and woodpeckers, lay white eggs, which are more easily seen in poor light.

The patterns of an egg are not simply a superficial coloring on the outside of the egg: some markings, such as dark spots and blotches, often concentrated toward the broader end of the egg, are integral parts of the shell's structure. The extra pigment in these blotches strengthens weak areas in the thinnest parts of the shell and they may have other functions as yet unknown.

Eggs are incubated by one or both parents.

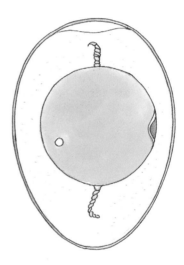

PARTS OF AN EGG

The egg yolk is held in place within the white, or albumen; at the broad end of the egg is an air cell.

An adult bird sits on the eggs to warm them until they hatch, usually with the aid of an "incubation patch"—bare skin on the belly of the incubating bird with no feathers but expanded blood vessels to supply extra heat. The developing chick receives nutrients from the yolk inside the egg and eventually breaks through the shell, using a tiny "tooth" at the tip of its bill.

Chicks can be described as either nidicolous (altricial) or nidifugous (precocial). Nidicolous chicks are weak, more or less naked, and blind,

NIDICOLOUS CHICK

Many chicks hatch out tiny, blind, and naked and then develop within the nest for several days.

NIDIFUGOUS CHICK

Some chicks hatch with a downy coat, strong legs, and good vision and leave the nest within hours.

and they remain in the nest, needing care and shelter until they grow a covering of feathers and are ready to fly. Nidifugous chicks are covered in warm down, are bright-eyed and active, leave the nest within a matter of hours, and immediately find their own food.

Most songbirds are nidicolous, while most terns, sandpipers, plovers, and chicken-like birds are nidifugous. Many species have chicks that are somewhat intermediate to these. The nestlings of hawks, owls, and herons, for example, are downy and more developed at

hatching than those of most songbirds. They are somewhat active when hatched and soon move about onto branches of their nest tree. Some, such as gull chicks, leave the nest in a downy covering, but require more care and feeding than typical nidifugous chicks.

All chicks need to be cared for to some extent, and they are brooded by a parent when they are cold and wet—the brooding adult typically calls the chicks, which nestle among the adult breast and belly feathers or under a drooped wing for warmth and protection.

HOW MANY EGGS?

A full set of eggs is called a clutch. Some long-lived birds that have a long period of immaturity, such as fulmars, lay a single egg each year. Small songbirds may lay one clutch of 10 to 12 eggs, their hatching coinciding with the peak availability of suitable food. Others produce two, three, or even four clutches of 3 to 5 eggs each during a season. Gamebirds, such as quail and pheasants, have only one clutch each year, but may lay up to 15 eggs; sometimes eggs are laid in the same nest by more than one female.

Rearing a family

Having put so much effort into finding and defending a territory, finding a mate, building a nest, and producing eggs, a parent bird must ensure that its young survive and thrive. To do so, it must defend them from predators, shelter them from hot sun, cold wind, or rain, and keep them well fed.

Chicks on the ground, such as those of plovers, seem extremely vulnerable, but they can scatter when danger threatens. Plover chicks also have excellent camouflage and quickly learn to "freeze," crouching stock still until a parent calls to say that danger has passed. In many ways, songbird chicks in a nest are more vulnerable. The nest may be well hidden and provide shelter, but the chicks are unable to move. Once they are found, there is no escape.

Defending the young

Even small birds fight against hawks, cats, and foxes as best they can. Birds as diverse as swallows, mockingbirds, terns, hawks, and owls may dive at the head of a person who comes too close to a nest containing chicks. Other birds, such as many sparrows, discreetly leave their nest when danger threatens: it is better for the parent to survive than to stay to face a human intruder and risk death, and the nest is less likely to be found if the parent bird is absent.

Even doves, particularly unsuited to aggressive attacks, have been seen diving at a small hawk that has caught a chick. Skuas are much more dynamic, aggressive birds, and they will dive at the head of a person near their breeding area, sometimes striking them. Female harriers, and much smaller Arctic Terns, practice similar behavior.

Feeding the family

Small birds such as tits visit the nest hundreds of times each day to bring food for a growing family. Chicks require a constant supply of energy and need brooding in bad weather—the parent shelters them under its wings and fluffed-out body feathers. Swifts feed their young on flying insects—not as reliable a food source as caterpillars and insects found on foliage. If the weather is wet and windy, the swifts may find it difficult to find food. At such times their chicks become temporarily torpid (they lower their body temperature and become inactive, thus conserving energy)—but swifts may fly hundreds of miles to avoid bad weather, or to exploit insects concentrated along weather fronts. They are among the finest meteorologists in the bird world.

In several species—as varied as Common Moorhens, Long-tailed Tits, Florida Scrub Jays, and Red-cockaded Woodpeckers—parent birds receive help to feed their young from other birds. In Common Moorhens, the young from an early brood help to feed chicks of a later brood. In Long-tailed Tits, helpers are usually brothers of the male parent, with no chicks of their own. In Florida Scrub Jays and Red-cockaded Woodpeckers, the helpers are usually offspring from previous years.

FEEDING CHICKS

Tiny chicks call loudly as long as they are hungry. This forces the parents to either keep feeding them or risk losing the results of weeks of intensive effort.

ENVIRONMENTAL EFFECTS

Some seabirds have struggled in recent years to find food for their chicks. Rising sea temperatures have affected the distribution and abundance of plankton, which are eaten by sand eels, the favored prey of some seabirds. Kittiwakes, terns, and guillemots have been seen catching dozens of tough, leathery pipefish, which have replaced the sand eels in some areas, and trying to feed them to their chicks. The chicks starve, surrounded by uneaten fish, which they are unable to swallow—many choke to death in the attempt. Global warming is also affecting the migratory patterns of some songbirds, which are returning earlier to their breeding grounds. In some cases, the young of these birds are hatching before the insects the parents need to gather for them are emerging, with catastrophic results. Nesting efforts fail for lack of food, and when subsequent insect emergence peaks, it remains uncontrolled by the lowered bird population.

CHAPTER TWO

Making the Most of Birding

This chapter contains expert tips, techniques, and advice on how best to observe birds: where, when, and what to look for, basic identification, taking notes, making sketches, and using binoculars. This section also teaches you more about the habitats and places you could venture to see more birds, as well information on transforming your backyard into a bird reserve.

Telling birds apart

Each species of bird is identifiably different in some way from other bird species. In many cases, the differences are obvious, but in others they are less so and may come down to extremely subtle variations in color, shape, or even the sounds a bird makes. The difficulty of identifying birds therefore varies from easy-for-beginners to challenging—even for specialists.

Beginners to bird-watching should start by identifying common garden and town birds. It is a good idea to get a few basic reference points—the commonest, simplest species— against which other birds can be compared. When you observe a "new" bird you can then think about how it looks compared with the birds you already know. Is it about the size of a sparrow, or as big as a pigeon? Is it the same shape as a robin, or similar to a starling but with a thicker bill? Such basic comparisons are invaluable and your bank of references will grow as you get to know more and more birds. You will then be able to make far more subtle and useful comparisons.

Noting size, shape, and behavior

The size of a bird is one of the first points you should note. This is not always easy, unless the bird is near another species or object with which you are familiar. It can be hard to judge the size of a bird that is flying high in the sky or perched far away across a field, but do your best. Try to get an idea of the bird's basic shape and its bulk. Is it a slim, lightweight, elegant bird, or a big, heavy, lumbering bird? Also take note of the proportions of its body parts in relation to each other. Does it have a large head with a long neck, bill, and legs? Or is the bird round-bodied, short-tailed, and short-legged? Try to get as much of an overall impression as you can when you spot a new or strange bird, including the way it moves and behaves. Is it walking quickly, hopping, or sitting still? Does it dash from bush to bush, or spend many minutes flying around over a field? If the bird is in a bush, does it sit quietly or is it constantly in motion, slipping through the foliage to hop from

GO PISH

There are several ways in which you can lure small songbirds to come closer so you can get a better look. "Pishing"—making a repeated, urgent "pshhh pshhh" sound—is one such method. This noise imitates that of a worried bird. Squeaking, with high-pitched squealing and squeaking sounds, also works well. You can also try playing the calls or songs of a bird on a CD to force a particular bird to approach to defend his territory against an intruder. This technique can work like magic, but it should be used sparingly and with caution. It tends to be disruptive and can be damaging for the bird. Use the playback technique with great restraint or not at all. Finally, some refuges and sanctuaries forbid the use of recorded calls and their use to lure an endangered species can be illegal.

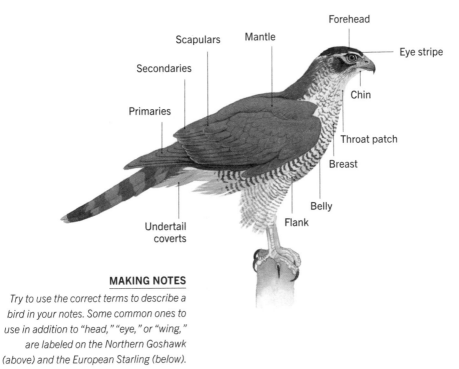

Forehead
Eye stripe
Chin
Throat patch
Breast
Belly
Flank
Undertail coverts
Primaries
Secondaries
Scapulars
Mantle

MAKING NOTES

Try to use the correct terms to describe a bird in your notes. Some common ones to use in addition to "head," "eye," or "wing," are labeled on the Northern Goshawk (above) and the European Starling (below).

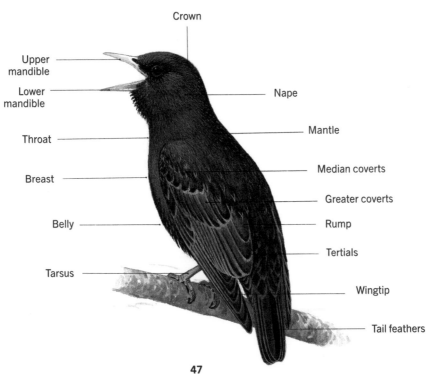

Crown
Upper mandible
Lower mandible
Throat
Breast
Belly
Tarsus
Nape
Mantle
Median coverts
Greater coverts
Rump
Tertials
Wingtip
Tail feathers

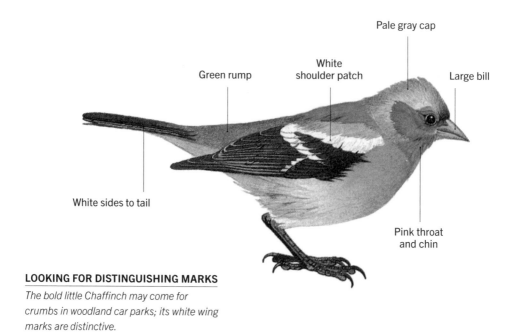

Pale gray cap

White shoulder patch

Green rump

Large bill

White sides to tail

Pink throat and chin

LOOKING FOR DISTINGUISHING MARKS

The bold little Chaffinch may come for crumbs in woodland car parks; its white wing marks are distinctive.

branch to branch? Does the bird sit and pick its food from the nearest twig or leaf, or does it hang upside down to reach around for its food? Is it alone or is it with other birds of the same or a different species?

Too often bird-watching beginners focus on just one or two points that are immediately obvious—such as a white patch on the body, or a yellow tinge to the wing—and ignore everything else, expecting that this will be enough to make an accurate identification. Usually, they will be disappointed. It is important to get as much information as you can. As you become more experienced, try to use the correct terms to describe the parts of a bird in your notes, including the tracts of feathers such as primaries and wing coverts. This will allow you to be more precise when describing what you have seen, and you will become fluent in the common language of bird-watchers.

Recognizing old friends

Eventually, the wealth of information you gather will allow you to identify common birds with ease. For example, sparrows feed on the ground and fly away fast and straight to the nearest bush if they are disturbed, while swallows hunt insects in the air and fly in a relaxed, flowing manner, rarely stopping to perch. Thrushes feed on the ground, hopping and running, and stopping to look and listen, or probe the ground with their bill. Goldfinches feed in little groups in the tops of thistles, flitting from flower to flower. Pigeons seem to shuffle along and sit on rooftops for minutes at a time, while kestrels sit on open posts and treetops, coming to the ground only to capture their prey. All of these impressions will soon become familiar and gradually you will be able to recognize what "kind" of bird you are looking at. You can then start to pay closer attention to finer details: observing colors and patterns in plumage and listening to bird sounds. Experienced bird-watchers often do not need these finer details to identify a bird. They will see a bird fly across a field or perch in a bush and be able to name it simply because "that is what it looks like." With some practice, you will be able to do the same.

Identifying birds becomes easier as they become more familiar. It is like spotting a friend on a busy street: you do not have to check the color of the eyes or the shape of the nose, you simply know who it is. However, if you are trying to find someone you have never seen before, and have just a photograph as a reference point, smaller details become more important. Bird-watching works in a similar way. At first you will be looking for all the specific markings of a bird illustrated in a field guide, but eventually you will be able to see the same bird fly by and identify it on sight.

USING YOUR EYES AND EARS

This adult male redpoll has a striking pink breast, but juveniles lack pink, and you need to look carefully for the black chin. Their calls can also help identify them.

Red cap

Tiny bill

Black chin

Buff wingbar

Streaked flank

Plain brown tail

Taking notes and making sketches

A notebook used to be part of every bird-watcher's outdoor kit. Now most bird-watchers use a digital camera fixed to a telescope to record what they see. Nevertheless, a notebook is exceptionally useful and is an ideal way to remember details about the birds you watch.

If you see something unusual when bird-watching, try to describe it in your notebook as thoroughly as possible. It is difficult to remember everything by memory alone, especially if you want to look up a bird in a book later to identify what you saw. Did the bird have pale brown legs and a buff band on the tail, or did you only read these things in the book? Were there four white spots on the wingtip or just three? It quickly becomes impossible to remember.

Take good notes

If possible, write everything down while you are still watching the bird. You can scribble down your own shorthand and expand on it later. If you get details wrong you can change them as you watch the bird and get a better view. You might write down "black bill" and later change it to "dark brown bill with paler base," or "eyes looked dark" might become "brown eyes with paler ring" once you see the bird more closely.

MALLARD

Even on a common bird like this, the pattern is hard to describe without at least a few specialist terms to pinpoint colors, but they are easy to learn.

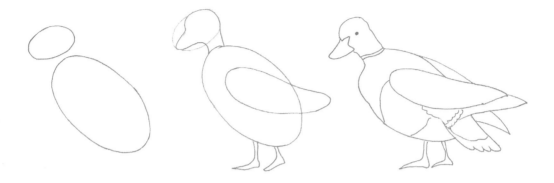

Do the best you can, but aim for as detailed an observation as possible.

You never know what crucial feature will help you identify a bird later, so make sure you write everything down. It can be helpful to make a sketch and label all the bird's features. This will make it clear what details are missing—perhaps you haven't labeled the legs, and this will prompt you to get a better look. Making a sketch forces you to look—closely, and properly. You can't draw the right pattern of streaks if you haven't seen them; you're not likely to get the head pattern right if you don't specifically choose to note head-pattern characteristics. Approach bird identification systematically, and know bird anatomy with an awareness of potential patterns of plumage and shapes of head, bill, wings, and tail.

If you are faced with a complicated bird, you can work at studying the details, noting them down, and putting them in a sketch, however crude it might be. Not only will you have to look at the bird carefully to get it right, but you will remember so much more about it, even months later. Taking a bird's picture gives you a quick reference, but fixes little in your mind. A photograph is a record of a tiny moment: it may not be as "true" as you think. How often

SKETCHING A DUCK
Start with a rough idea of shapes and proportions, then add shape, then colors and patterns.

do you take a photograph of someone and find it doesn't really look right—perhaps they moved, or blinked, or turned away at the crucial moment. A photograph of a bird can be just as inaccurate. A drawing, or a written description, made as the bird moves and feeds, flies, and preens, will be a much more complete record than a moment caught by the camera. Of course, you may do both, but don't neglect the old-fashioned notebook and drawing.

How to make a sketch
Even if you are not a skilled artist, any drawing is better than none. Use rough egg shapes to get the basic idea of a head and body—as near as you can get to the right proportions. Is the body elongated horizontally, or vertically? Is it round or thin? Add wings, a tail, legs, and a bill as well as you can. Now put labels on all the features, making short notes with lines pointing to the position on the sketch. Make more sketches as you get a better view of the bird or see it from different angles.

TAKE NOTES RIGHT AWAY

Remember, once you have read a description of a bird, or looked at a photograph or a painting, or listened to a sound track, it is more difficult to remember exactly what you originally saw or heard—details may become obscured. Once you have gone home, you cannot look at the live bird again to check something—so take notes and make sketches while you can. You will not regret it.

Sounds

Bird calls and songs are crucial in helping you to identify a species. Bird-watchers also use them to locate birds, many finding more birds by using their ears than by using their eyes. You can buy CDs containing recordings of bird sounds, but learning them directly from the bird that makes them is more effective, as it puts the sound into context and makes it more memorable. The best situation is to see the bird actually making the noise; then you know for certain what noise a species makes and are more likely to remember it later.

You may find it helpful to write down what a bird call sounds like. Use simple words or syllables to describe how the call sounds to you, such as tchew, twit, sweep, or chirrup. Add descriptions of the noise, such as liquid, metallic, sharp, hard, soft, silky, or abrupt. Many bird calls are difficult to describe because they consist of an extended vowel sound without a hard consonant—but do your best.

You can add a clue to the rise and fall and inflection of a call by drawing a line above it. For example, you can use a thick line that tapers out to indicate a call that fades away, or make the line bend up or down to show the rise and fall in pitch of the call. If you need to refer to a book or CD when you get home, these efforts can help you to remember what you heard.

Keeping a diary

Most bird-watchers like to keep some sort of log of the birds they have seen. Those who began bird-watching before personal computers were commonly used probably kept a simple written diary, a card index, or perhaps merely a life-list of birds they have seen. Many people who are starting out today will prefer to keep computerized records. However, there are pros and cons for every system and none, save perhaps for a sophisticated computer program, is perfect.

A daily bird-watching diary is useful. Important things to note include the date of the observations, the time, your location, perhaps the weather, and a list of birds you see. You can liven your diary up with the names of your companions, or note the day you first used your new binoculars. A good diary is far more than a basic list of birds—it is also a personal story that can include your thoughts, observations, and excitement at the best (or worst) of times.

In some special circumstances (if, for example, you are keeping detailed reports at a nature reserve), you might note everything you see. Usually, though, this becomes tedious and repetitive, and you will instead be more selective. By neglecting details, however, you risk losing good data on birds that are common now, but might not always be so. There has to be a compromise: keeping notes should add enjoyment to your hobby, not make it a chore, but it should also be useful.

The more you bird-watch, the more you will learn what is worth writing down and what is not. You will, after all, be keeping these notes for yourself, as a rule. They will be private rather than public records. You should send in all the significant and useful notes to a local or county bird recorder, to add to the official record. To do this easily, you may need a system that allows you to find all your notes from one location

or all your notes on a particular species. This is where a card index works better than a daily diary. A computer log, on which you can search for specific words, is obviously the most efficient option. While a card index is helpful for finding all your sightings of any one species, it fails to tell you what you saw at a particular place or on a specific day, and it does not bring back memories, or record the atmosphere, in the way that a daily log does. And for many people, a computer diary isn't as "friendly" as a paper one.

THE PROS AND CONS

Do you want to keep the notebooks you take out bird-watching with you? Or do you prefer a more permanent record, written up in an exercise book, or typed onto a computer? You can, of course, keep both, but space becomes a problem over the years. In the end, you will find your own preferred method, but think about what you want from a notebook and try to make it practical—get it to work for you.

	PROS	CONS
Notebook	Portable	Takes up space
	User-friendly	Can be difficult to find information
	Personal and evocative	
Desktop computer	Quick, easy searches	Not portable
	Compact	Cannot use it in the field
	Programs can sort information	
Mobile phone	Quick, easy searches	Impersonal and unevocative
	Compact and portable	Dependent on battery life
	Programs can sort information	

Choosing and using binoculars

Apart from a field guide illustrating the area's local birds, the only equipment a bird-watcher needs is a pair of binoculars. A good pair will last a lifetime, so choose carefully and get the best you can afford.

Recommending a particular make and model of binoculars is not easy, because different shapes and sizes suit different people and manufacturers are constantly improving their products and coming out with new ones. For bird-watching, a small, lightweight pair is ideal, but above all, it is important to make sure that binoculars are comfortable to use. They should "fit" your hands and the region around the eyes, so comfort largely depends on personal preference and the size and shape of your hands and face.

Types of binoculars

Binoculars come in two basic types: the roof prism, which looks straight-sided and slim, and

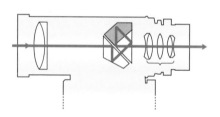

ROOF PRISM
This design is the more modern and efficient type of binoculars.

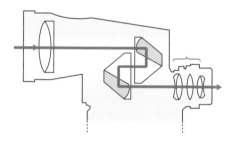

PORRO PRISM
Light is reflected through a prism before reaching the eyepiece.

KEEP THEM CLEAN!

Eyepieces are often smeared with sunscreen, insect repellent, or makeup, and the larger lenses become dirty. Insect repellent can also damage the coating of lenses. When cleaning, avoid rubbing because it damages the lens coating and may grind in dirt, scratching the lens itself. Blow away dirt and dust, then gently brush the lens with a soft cloth designed for lens cleaning. Do not use a shirt or handkerchief, or any tissue paper other than specially made lens paper. It is always best to use a liquid lens cleaner with your lens cloth.

the porro prism, which looks "step-sided" and has much wider front lenses (objective lenses) than eyepieces (ocular lenses). The roof prism is the more efficient system, but you may not notice a difference in the images it magnifies. They are also smaller and lighter in weight, and you can get compact or miniature models, but smaller binoculars let in less light and are less useful in low-light conditions. Try both types at a camera or optics store before making a decision.

One thing to keep in mind: porro prism binoculars have an external focusing mechanism—the front lenses move in and out on a greased rod when the viewer focuses on something. That grease is exposed to dust in the air. Over time, the grease becomes saturated with dust and dirt, making focusing more difficult or even impossible. Roof prism binoculars, on the other hand, have an internal focusing mechanism and therefore do not have any external moving parts that will become gummed up with grease.

Power and lens size

The magnification level (the number of times that an object will be enlarged) and the diameter of the larger lenses (in millimeters) of binoculars is always specified: 10x40 binoculars, for example, magnify the object by 10 times and have 40 mm lenses. For bird-watchers, a magnification, or "power," between 7 and 10 is generally suitable. However, the higher the power, the more difficult it will be to hold them steady—every movement of the hand is magnified as much as the image. You will need a bright image if you are to accurately identify birds. Larger front lenses allow more light in, and the higher the binoculars' power, the larger the lenses need to be to show the magnified picture brightly.

When choosing binoculars, divide the diameter of the front lenses (the second number) by the magnification (the first number) to get the diameter of the "exit pupil"—the bright beam of light that exits the eyepiece. There is no point to having an exit pupil that is larger than the pupil of your eye, as any extra light will be wasted. At its largest, your pupil will be about 7 mm, but in sunlight it is much smaller.

Focusing binoculars

The first adjustment to make takes account of the distance between your eyes, which varies greatly in different people. First, push the barrels closely together. Then point them toward a distant object, raise them to your eyes, and slowly increase the space between the barrels until you see a maximum field of view. This happens when the individual fields of view of the two barrels (the separate images seen by your two eyes) appear to merge and form an oval or circle. The binoculars are now adjusted to fit the distance between your eyes.

Next you must focus the two barrels separately, to take account of the variations in sight between your eyes—even people with good vision usually do not have equally matched sight in both eyes. Once you have made these adjustments, your binoculars are ready to use for observation.

Practice using your binoculars until you can raise them to your eyes without moving your head—this will enable you to keep the bird in sight and avoid disturbing it. Looking directly at the bird (not at the binoculars), raise the binoculars smoothly to your eyes. Swinging your head from side to side with your binoculars in place to get the bird in view is not the way to do it. You can practice this exercise using any object.

Farther afield

Bird-watching from your window, looking at garden birds, is a good way to start, and garden birds will keep you amused and fascinated for a lifetime. But no doubt you will soon want to explore farther and see different kinds of birds. Your daily routine will reveal only a tiny fraction of the birds out there. Birds are everywhere and you will want to know what species you are looking at.

Look through a field guide and you will see birds you have hardly dreamed of, and they may be close by. If you want to see new birds, the first thing to do is to head for water. Try the beach, a flooded gravel pit, or a reservoir. Water and the water's edge provide habitats for a huge number and variety of birds.

CANADA GOOSE

Even common birds such as the Canada Goose have particular needs and take you away from your backyard—head for water to see this one.

Watching around water

Gravel pits tend to be quite deep and steep-sided, so they are ideal for several kinds of ducks, especially diving ducks such as Ring-necked Ducks, Redheads, and Buffleheads, but not necessarily for those that potter about in the shallows such as Green-winged Teal, and shovelers. Gravel pits often contain grebes and perhaps geese or swans. Reservoirs in steep-sided valleys between high hills tend to be cold, deep, and acidic, and have few birds on the whole. But those created on lowland farmland in wide, shallow valleys can be exceptional. The shores become muddy, and any small drop in water level is likely to reveal vast areas of mud, which becomes overgrown in the summer but flooded again in fall, releasing a superabundance of seeds for winter wildfowl to eat.

The shallow sides attract wading birds on their spring and fall migrations, too. These can be some of the most exciting birds you can find—not only are the common ones beautiful species that may be on their way to the Southern Hemisphere from the Arctic, but the chances of finding something quite rare are fairly good. The water's edge is excellent, too, for birds such as meadowlarks, pipits, finches, and various sparrows. Insects over the water attract early migrants, and flocks of swallows and swifts often gather there in bad weather, when food is hard to find elsewhere. Over the water, too, may be migrant terns, and from late summer to spring, there may well be a big nighttime roost of gulls flying in to find a safe place to sleep after feeding on local farmland or refuse tips.

Any area around these watery places is worth exploring. Willow thickets, bramble and scrub patches, reeds, and rushes are all likely to have interesting breeding birds and migrants that rely on the proximity to water to provide a reliable supply of insects.

On the coast the possibilities are endless. A rocky shore appeals to certain species, while a sandy one may be less rich in birds. But mud, especially in a big estuary, is a super-rich habitat with a fantastic supply of food, which is refreshed twice a day by the tide. As the tide falls, huge numbers of shorebirds of many kinds spread out over the flats. This can be challenging bird-watching, as birds are at long range and conditions are often difficult, but it can be remarkably rewarding. Often, too, there are roads—even urban promenades—that overlook wonderful places for birds in estuarine locations. A trip to the seashore to look for birds should be timed to arrive at low tide.

Other habitats

Woodlands are full of birds, but are often tricky places in which to bird-watch. Dense summer foliage makes things difficult and, in winter, birds are fewer and more concentrated in wandering flocks. It is best to go early in the day in spring, when everything is singing and the leaves are still sparse, or to wander patiently in winter until you come across a mixed flock, when things will suddenly be hectic for a minute or two before they pass by. For birds of prey, it is often best to stay outside the wood, preferably on a hill from which you can look out across the forest and watch for birds displaying

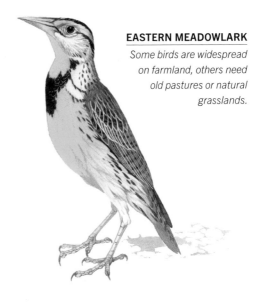

EASTERN MEADOWLARK
Some birds are widespread on farmland, others need old pastures or natural grasslands.

over the forest in spring. Woodpeckers, warblers, thrushes, nuthatches, and creepers are typical woodland birds.

Scrublands are wide-open, sometimes bleak places, and birds can be few and far between except in spring and summer. You need to be patient. Walk slowly and carefully, checking bushes and thickets and looking overhead for passing birds of prey. Their special species make these habitats well worth the visit, but finding all their secrets may take time.

Mountains and hillsides have some special birds, but these are expansive habitats, and you must be sure you are safe—the weather can change dramatically. Take care, wear the right footwear, take food and drink, a map, and warm clothing. But do try to see some mountain birds if you can. It is often possible to do so from a roadside, but you might need to climb a hilltop or explore a valley: keep high and look over the lower areas. The Golden Eagle, Peregrine Falcon, and ptarmigans are highlights to seek out in the mountains.

Make your backyard a bird reserve

Transforming your yard into a refuge for wildlife is easy to do, is good fun, and brings great rewards. With a little effort, you can attract beautiful songbirds and other interesting wildlife to your yard, which the whole family can enjoy watching. Creating a bird sanctuary can help you unwind and relieve the stress of the working week.

Gardening practices that help wildlife, reduce chemicals in your yard, and conserve water also help to improve the quality of air, water, and soil throughout your neighborhood. Backyards can support wildlife all year round, but birds need your help most during the colder months and in the early spring when wild food is still scarce and winter supplies are already exhausted. Feeding wild birds is a popular hobby, second only to gardening in North America.

A safe environment
Keeping your bird feeders clean is vitally important—unhygienic practices are more likely to lead to disease in the birds that feed in your garden. Placing them in the right place is also necessary if you want to keep birds safe from predators and other dangers (see How to Feed Birds in the Garden, pp. 28–29). Predation by household cats is a major cause of backyard bird deaths, and a study by the American Birding Association found cats to be significant killers of birds that come to feeders. A single domestic household cat can kill more than 100 birds and small mammals each year: even if one bird a month is killed in this way, it still translates to millions of birds being killed by pet cats throughout the U.S. Make sure you place feeders away from anywhere a cat can lie in ambush.

Window strikes are a frequent cause of injury or death to birds—reflections on a big window on a sunny day may create the illusion of an open space that birds may try to fly through. You should take measures to reduce the risk of window strikes. Simply relocating a feeder may help the problem. There are also many ways to disrupt reflections in windows, such as putting stickers on the windows or hanging netting or objects outside them. Shiny objects in particular are more likely to deter birds from approaching.

Making your garden hospitable
What kind of habitat do you have in your own backyard? A pile of brush may be unsightly, but it can help birds escape predators such as cats and Sharp-shinned Hawks. A good thick shrub or hedge can do the same. It should be within easy reach of the feeder, but not so close that a cat can leap out to pounce on a bird.

You should provide birds with water for drinking and bathing. If you do not have room for a pond, a birdbath will suffice, but make sure you keep it topped up with clean, fresh water. A small pond is by far a better option, though, if you are able to create one. It should be quite deep, and lined with old carpet beneath a layer of plastic or rubber pond liner. Also add a layer of earth. Or else you can simply buy an artificial pond. Make sure some of the pond's edges are shallow and slope gradually so that birds can walk in gently: they don't like having to plunge straight in out of their depth. Add a selection of native waterside plants, but leave some open space, too, around the pond edge. It is best not to have fish in a garden pond, as they eat insects and tadpoles. This setup is better suited for a wildlife garden.

As well as setting up a selection of feeders, you should provide other food sources for birds: plants and trees with fruits and berries, or flowers that attract plenty of insects. If you have the space, dead trees or branches help birds such as woodpeckers to find food and nest sites.

If you want to get serious about backyard wildlife, you can register with the Backyard Wildlife Habitat Program with the National Wildlife Federation, which acknowledges the efforts of people who garden for wildlife, gives personal registration certificates, and adds wildlife gardens to the national register of backyard wildlife habitats.

BIRDBATH
Birds need to bathe regularly to keep in good shape, even in cold weather.

Bird Catalog

The pages that follow feature over 50 common birds that can be found in the United States. Those that can be seen in and around gardens and towns are examined, along with some that are easy to see up close in locations such as a town park (especially if it has a lake) or in nearby farmland. This catalog includes illustrations and details that will help you to identify these species: size, habitat, distribution during the breeding season and during winter, and preferred foods. Three of the species here—the Eurasian Starling, House Sparrow, and Rock Pigeon—are non-native species that were introduced to North America. Once you go beyond the habitats described here, you will begin to see many more species, and a more comprehensive identification guide will be needed.

27 in.
69 cm

DOUBLE-CRESTED CORMORANT

SIZE: 27 inches (69 cm)

HABITAT: Rivers, lakes, park lakes, coasts

BREEDS: Most of North America

WINTERS: Pacific coast, Atlantic and Gulf coasts, North Carolina to Belize

FOOD: Fish

A*s*

KEY:  Male   Female   Adult winter   Adult summer

MALLARD

SIZE: 16 inches (41 cm)

HABITAT: Water, especially freshwater, waterside areas

BREEDS: Almost all of North America

WINTERS: Northern birds move south

FOOD: Seeds, shoots, aquatic vegetation, invertebrates

 ♂

16 in.
41 cm

AMERICAN COOT

SIZE: 12 inches (30 cm)

HABITAT: Lakes, reservoirs, rivers, parks, ponds

BREEDS: Canada to Ecuador

WINTERS: Resident in breeding areas

FOOD: Aquatic plants and invertebrates

Aw As

12 in.
30 cm

 J Juvenile

 Feeds at birdtables

 Feeds at hanging feeders

 Uses nest boxes

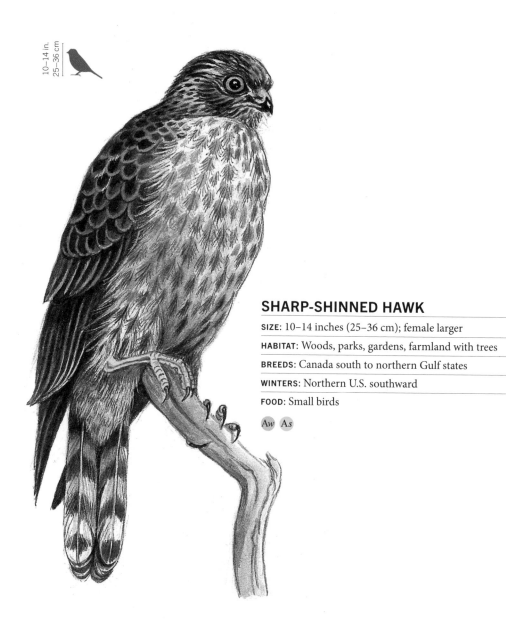

10–14 in.
25–36 cm

SHARP-SHINNED HAWK

SIZE: 10–14 inches (25–36 cm); female larger

HABITAT: Woods, parks, gardens, farmland with trees

BREEDS: Canada south to northern Gulf states

WINTERS: Northern U.S. southward

FOOD: Small birds

Aw As

KEY: ♂ Male ♀ Female   Aw Adult winter   As Adult summer

AMERICAN KESTREL 🏠

SIZE: 8½ inches (22 cm)

HABITAT: Parks, gardens, open fields, bushy areas, woodland edge

BREEDS: Most of North and South America

WINTERS: Resident in breeding areas

FOOD: Small rodents, insects, small birds

Aw As

8½ in.
22 cm

♀

♂

RING-BILLED GULL

SIZE: 16 inches (41 cm)

HABITAT: Lakes, quays and beaches, farmland

BREEDS: Canada, northern U.S.

WINTERS: Coasts, southern U.S. to Gulf, Mexico, Cuba

FOOD: Fish, scraps, worms

As

16 in.
41 cm

J Juvenile | 🪧 Feeds at birdtables | Feeds at hanging feeders | 🏠 Uses nest boxes

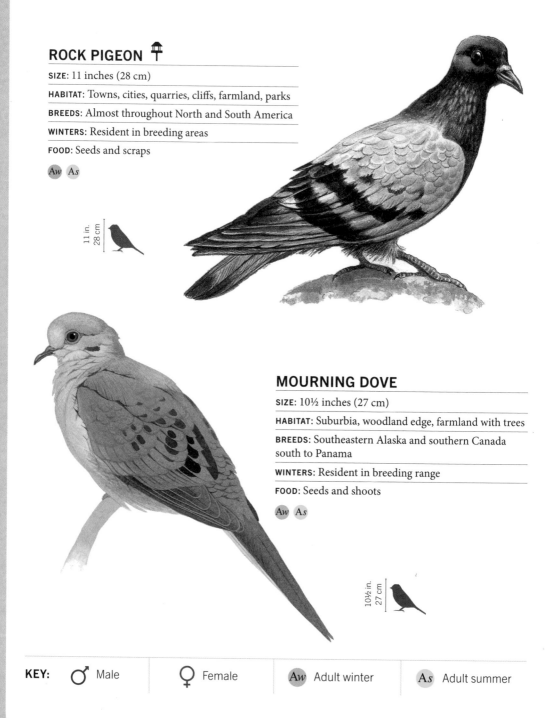

ROCK PIGEON

SIZE: 11 inches (28 cm)

HABITAT: Towns, cities, quarries, cliffs, farmland, parks

BREEDS: Almost throughout North and South America

WINTERS: Resident in breeding areas

FOOD: Seeds and scraps

Aw As

11 in.
28 cm

MOURNING DOVE

SIZE: 10½ inches (27 cm)

HABITAT: Suburbia, woodland edge, farmland with trees

BREEDS: Southeastern Alaska and southern Canada south to Panama

WINTERS: Resident in breeding range

FOOD: Seeds and shoots

Aw As

10½ in.
27 cm

KEY: ♂ Male ♀ Female Aw Adult winter As Adult summer

11 in.
28 cm

YELLOW-BILLED CUCKOO

SIZE: 11 inches (28 cm)

HABITAT: Open woods, orchards, streamside trees

BREEDS: Almost throughout U.S.; rare in northwestern quarter

WINTERS: South America

FOOD: Insects, especially caterpillars

A*s*

 Juvenile

 Feeds at birdtables

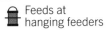 Feeds at hanging feeders

 Uses nest boxes

GREAT HORNED OWL

SIZE: 20 inches (51 cm)

HABITAT: Forests, woodlands, thickets, open country with scattered trees

BREEDS: Alaska south throughout Americas

WINTERS: Resident in breeding range

FOOD: Rodents and larger mammals, birds

20 in.
51 cm

KEY: Male | Female | Adult winter | Adult summer

CHIMNEY SWIFT

SIZE: 5 inches (13 cm)

HABITAT: Open air throughout, nesting in hollow trees and chimneys

BREEDS: Eastern North America from southern Canada to Mexico

WINTERS: South America

FOOD: Flying insects

A*s*

5 in.
13 cm

RUBY-THROATED HUMMINGBIRD 🏮

SIZE: 3 inches (7½ cm)

HABITAT: Gardens, woodland edges with flowers

BREEDS: Eastern North America from southern Canada to Gulf states

WINTERS: Mexico and Central America

FOOD: Nectar, tiny insects, spiders

A*s* ♂

3 in.
7½ cm

 Juvenile

 Feeds at birdtables

 Feeds at hanging feeders

 Uses nest boxes

RED-BELLIED WOODPECKER

SIZE: 8½ inches (22 cm)

HABITAT: Woodlands, orchards, gardens, parks

BREEDS: Great Lakes; New England to Gulf states

WINTERS: Resident in breeding areas

FOOD: Insects, seeds, berries

Aw As ♂

8½ in.
22 cm

*Red-shafted
form (rear);
Yellow-shafted
form (front)*

NORTHERN FLICKER

SIZE: 11 inches (28 cm)

HABITAT: Open forests, farmland with trees, parks, towns

BREEDS: Alaska south through Canada and U.S.

WINTERS: Resident in breeding areas

FOOD: Insects, seeds

Aw As ♂

11 in.
28 cm

KEY: ♂ Male   ♀ Female   Aw Adult winter   As Adult summer

7½ in. / 19 cm

RED-HEADED WOODPECKER

SIZE: 7½ inches (19 cm)

HABITAT: Farms, roadsides, open woodland, towns

BREEDS: Eastern North America from southern Canada to Gulf states

WINTERS: Northern birds move south

FOOD: Insects, fruit, some seeds

Aw As

 Juvenile

 Feeds at birdtables

 Feeds at hanging feeders

Uses nest boxes

7½ in.
19 cm

♂

♀

HAIRY WOODPECKER

SIZE: 7½ inches (19 cm)

HABITAT: Woods, parks, orchards, suburbia

BREEDS: Alaska and forested Canada south to western Panama

WINTERS: Resident in breeding range

FOOD: Insects, berries, seeds, nuts

A*w* A*s*

KEY: ♂ Male   ♀ Female   A*w* Adult winter   A*s* Adult summer

5 in.
13 cm

TREE SWALLOW

SIZE: 5 inches (13 cm)

HABITAT: Open areas near water, meadows, lakes

BREEDS: Alaska, Canada, much of eastern U.S.

WINTERS: Southern U.S. to Central America

FOOD: Flying insects

A*s*

BARN SWALLOW

SIZE: 6 inches (15 cm)

HABITAT: Farms, streamsides, meadows, parks

BREEDS: Most of North America

WINTERS: Costa Rica to Argentina

FOOD: Flying insects

A*s*

6 in.
15 cm

 J Juvenile

 Feeds at birdtables

 Feeds at hanging feeders

 Uses nest boxes

7 in.
18 cm

PURPLE MARTIN

SIZE: 7 inches (18 cm)

HABITAT: Open country, farms, gardens; breeds in artificial martin houses

BREEDS: Southern Canada to Gulf states; less common in western regions

WINTERS: South America

FOOD: Flying insects

A*s*

KEY: ♂ Male   ♀ Female   A*w* Adult winter   A*s* Adult summer

10 in.
25 cm

BLUE JAY

SIZE: 10 inches (25 cm)

HABITAT: Oak and pine woods, suburbia

BREEDS: Eastern North America from southern Canada to Gulf states

WINTERS: Resident in breeding range

FOOD: Insects, seeds, nuts, berries, eggs, scraps

Aw As

CALIFORNIA SCRUB-JAY

SIZE: 13 inches (33 cm)

HABITAT: Woodlands, chaparral, pastures, backyards

BREEDS: Western U.S. and south to central Mexico

WINTERS: Resident in breeding areas

FOOD: Insects, nuts, seeds, berries, rodents, eggs

Aw As

13 in.
33 cm

| Juvenile | Feeds at birdtables |  Feeds at hanging feeders | Uses nest boxes |

AMERICAN CROW

SIZE: 21½ inches (55 cm)

HABITAT: Woods, parks, farmlands

BREEDS: North America, from south Canada to New Mexico

WINTERS: Mostly resident in breeding areas

FOOD: Omnivorous

Aw As

BLACK-CAPPED CHICKADEE

SIZE: 4½ inches (11.5 cm)

HABITAT: Mixed and deciduous woods, thickets, gardens

BREEDS: Alaska, central and southern Canada, northern half of U.S.

WINTERS: Resident, some southward movement in hard winters

FOOD: Insects, seeds

Aw As

KEY: ♂ Male   ♀ Female   Aw Adult winter   As Adult summer

5¼ in.
14 cm

OAK TITMOUSE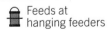

SIZE: 5¼ inches (14 cm)

HABITAT: Conifers and deciduous woodlands, and scrub

BREEDS: California and Mexico; just reaching Oregon

WINTERS: Resident in breeding areas

FOOD: Small insects, seeds, nuts, berries

A*w* A*s*

 Juvenile | Feeds at birdtables | Feeds at hanging feeders | Uses nest boxes

5–6 in.
13–15 cm

WHITE-BREASTED NUTHATCH

SIZE: 5–6 inches (13–15 cm)

HABITAT: Forests, shelter belts, gardens with trees

BREEDS: Southern Canada to Mexico

WINTERS: Resident in breeding range

FOOD: Nuts, berries, seeds, insects

Aw As

KEY: ♂ Male   ♀ Female   Adult winter   Adult summer

HOUSE WREN

SIZE: 4–5 inches (10–13 cm)

HABITAT: Woods, thickets, gardens, parks

BREEDS: Southern Canada to Argentina

WINTERS: Northern birds move into southern parts of range

FOOD: Insects, spiders, small seeds

A*w* A*s*

4–5 in.
10–13 cm

WINTER WREN

SIZE: 3½–4 inches (9–10 cm)

HABITAT: Woods, gardens, parks, thickets

BREEDS: Southern Canada, Pacific states, northern U.S. east of Rocky Mountains

WINTERS: Canada from southeast Alaska, Pacific states, southern half of eastern U.S.

FOOD: Insects, spiders

A*w* A*s*

3½–4 in.
9–10 cm

J Juvenile | Feeds at birdtables | Feeds at hanging feeders | Uses nest boxes

GOLDEN-CROWNED KINGLET

SIZE: 3½ inches (9 cm)

HABITAT: Coniferous woods, gardens

BREEDS: Southern Alaska and Canada south to North Carolina

WINTERS: Southward to Gulf states

FOOD: Insects, spiders

 A*w* A*s*

3½ in.
9 cm

RUBY-CROWNED KINGLET

SIZE: 4–5 inches (10–13 cm)

HABITAT: Conifers and mixed woodlands

BREEDS: Alaska, Canada, western U.S., extreme north-eastern U.S.

WINTERS: To Gulf states, Central America

FOOD: Insects, spiders

A*w* A*s*

♀

♂

4–5 in.
10–13 cm

KEY: ♂ Male   ♀ Female   A*w* Adult winter   A*s* Adult summer

AMERICAN REDSTART

SIZE: 5 inches (13 cm)

HABITAT: Deciduous woods

BREEDS: Canada and eastern U.S.

WINTERS: Mexico, West Indies, Brazil

FOOD: Insects

A*s* ♀

 J Juvenile

 Feeds at birdtables

 Feeds at hanging feeders

 Uses nest boxes

WRENTIT

SIZE: 6 inches (15–16 cm)

HABITAT: Chaparral, confierous brushland, coastal scrub

BREEDS: Central and coastal California, north to coastal Oregon

WINTERS: Resident in breeding areas

FOOD: Insects and berries

6 in.
15–16 cm

KEY: ♂ Male   ♀ Female   Adult winter   Adult summer

YELLOW-RUMPED WARBLER

SIZE: 5–6 inches (13–15 cm)

HABITAT: Coniferous and mixed woods, thickets

BREEDS: Alaska, Canada, south to Mexico in west and Tennessee in east

WINTERS: Northeastern U.S., Pacific and Gulf coasts south to Panama

FOOD: Insects

As ♂

5–6 in.
13–15 cm

AMERICAN ROBIN

SIZE: 9–11 inches (23–28 cm)

HABITAT: Cities, towns, parks, gardens, open woodland, orchards

BREEDS: North America, rarely nests in north Florida

WINTERS: Across U.S. and Central America

FOOD: Worms, insects, berries, and fruit

As ♂

9–11 in.
23–28 cm

 Juvenile   Feeds at birdtables    Feeds at hanging feeders   Uses nest boxes

LOGGERHEAD SHRIKE

SIZE: 9 inches (23 cm)

HABITAT: Shrubs, farmland, desert, fields, parks

BREEDS: Southern, west central, and western U.S.

WINTERS: Mostly resident in breeding areas but also moving south

FOOD: Insects, birds and small mammals

Aw As

9 in.
23 cm

CEDAR WAXWING

SIZE: 6 inches (15 cm)

HABITAT: Woodland, orchards, parks, gardens

BREEDS: Alaska, Canada to central U.S.

WINTERS: From southern Canada south to Panama

FOOD: Insects, berries

Aw As ♂

6 in.
15 cm

KEY: ♂ Male   ♀ Female   Aw Adult winter   As Adult summer

8 in.
20 cm

♂

J

EUROPEAN STARLING

SIZE: 8 inches (20 cm)

HABITAT: Cities, parks, farms, gardens

BREEDS: Southern Canada and whole of U.S.

WINTERS: Across U.S., northern birds move south

FOOD: Insects, worms, fruit, berries, seeds

A.s

 Juvenile | Feeds at birdtables |  Feeds at hanging feeders | Uses nest boxes

FOX SPARROW

SIZE: 7 inches (18cm)

HABITAT: Conifer and mixed woods, thickets, and shrubs

BREEDS: Northern Canada and Alaska, and western Canada south in mountain to Nevada and Colorado

WINTERS: West coast down to Texas and North Carolina

FOOD: Insects, seeds and fruits

A*w* A*s*

SONG SPARROW

SIZE: 5½–7 inches (14–18 cm)

HABITAT: Thickets, brushy areas, old fields, gardens

BREEDS: Across most of North America

WINTERS: South from Nebraska to New Mexico, Texas, and Florida

FOOD: Seeds and small insects

A*w* A*s*

KEY: ♂ Male | ♀ Female | A*w* Adult winter | A*s* Adult summer

5¼–6 in.
14–17 cm

DARK-EYED JUNCO

SIZE: 5¼–6 inches (14–17 cm)

HABITAT: Conifers and mixed woods, gardens, thickets, open ground in winter

BREEDS: Alaska, Canada, south in mountains to Arizona in the west, Georgia in the east

WINTERS: South to Gulf states, Mexico

FOOD: Insects, seeds, berries

A*s* ♀

 Juvenile | Feeds at birdtables | 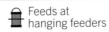 Feeds at hanging feeders | Uses nest boxes

6 in.
15 cm

WHITE-CROWNED SPARROW

SIZE: 6 inches (15 cm)

HABITAT: Bushy places, roadsides, woodland edges

BREEDS: Alaska, Canada, western U.S.

WINTERS: South to Gulf states, Mexico, Cuba

FOOD: Seeds, insects

A*s*

KEY: Male   Female   A*w* Adult winter   A*s* Adult summer

5–6 in.
13–15 cm

HOUSE SPARROW

SIZE: 5–6 inches (13–15 cm)

HABITAT: Gardens, parks, suburbs, farms

BREEDS: Almost all of North America

WINTERS: Resident in breeding areas

FOOD: Insects, seeds, scraps

A*s*

♀

♂

7 in.
18 cm

BROWN-HEADED COWBIRD

SIZE: 7 inches (18 cm)

HABITAT: Farms, fields, roadsides, woodland edges

BREEDS: Southern Canada to Mexico, northern Florida

WINTERS: Northern birds move mainly to southern half of U.S.

FOOD: Insects, fruits, grain, seeds

A*w*   A*s*   ♂

| **J** Juvenile | Feeds at birdtables | Feeds at hanging feeders | Uses nest boxes |

COMMON GRACKLE

SIZE: 10–12 inches (25–30 cm)

HABITAT: Farms, towns, woodland edge

BREEDS: Central Canada through eastern U.S.

WINTERS: Southern two-thirds of U.S. east of Rockies

FOOD: Insects, invertebrates, fruit, berries

Aw As ♂

10–12 in.
25–30 cm

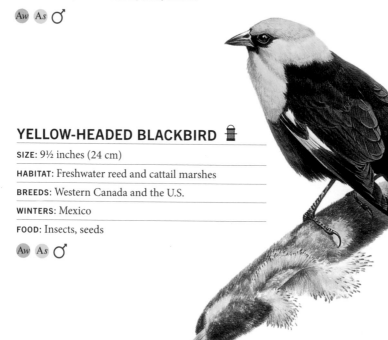

YELLOW-HEADED BLACKBIRD

SIZE: 9½ inches (24 cm)

HABITAT: Freshwater reed and cattail marshes

BREEDS: Western Canada and the U.S.

WINTERS: Mexico

FOOD: Insects, seeds

Aw As ♂

9½ in.
24 cm

KEY:	♂ Male	♀ Female	Aw Adult winter	As Adult summer

8–9½ in.
20–24 cm

A*s* ♂

7–7½ in.
18–19 cm

A*w* A*s* ♀

RED-WINGED BLACKBIRD

SIZE: Male 8–9½ inches (20–24 cm);
Female 7–7½ inches (18–19 cm)

HABITAT: Marshes, watercourses, cultivated land,
wood edge

BREEDS: North and Central America

WINTERS: Mexico

FOOD: Insects, invertebrates, seeds

 Juvenile

 Feeds at
birdtables

 Feeds at
hanging feeders

Uses
nest boxes

PINE SISKIN

SIZE: 4–5 inches (10–13 cm)

HABITAT: Conifers, mixed woods, bushy places

BREEDS: Southern Canada south through U.S.

WINTERS: Southward to Mexico

FOOD: Seeds, insects

Aw ♀

4–5 in.
10–13 cm

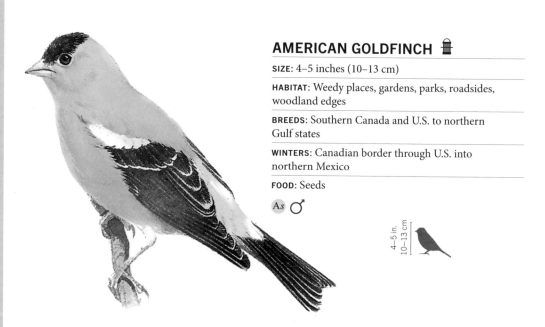

AMERICAN GOLDFINCH

SIZE: 4–5 inches (10–13 cm)

HABITAT: Weedy places, gardens, parks, roadsides, woodland edges

BREEDS: Southern Canada and U.S. to northern Gulf states

WINTERS: Canadian border through U.S. into northern Mexico

FOOD: Seeds

As ♂

4–5 in.
10–13 cm

KEY: ♂ Male | ♀ Female | Aw Adult winter | As Adult summer

5–6 in.
13–15 cm

PURPLE FINCH

SIZE: 5–6 inches (13–15 cm)

HABITAT: Woods, groves, suburbs in winter

BREEDS: Canada, northeastern and western U.S.

WINTERS: Widespread across U.S.

FOOD: Seeds

A*s*

HOUSE FINCH

SIZE: 5–6 inches (13–15 cm)

HABITAT: Cities, suburbs, gardens, farms

BREEDS: Widespread in western U.S., spreading locally in east

WINTERS: Resident in breeding areas

FOOD: Seeds

A*s* ♂

5–6 in.
13–15 cm

J Juvenile | Feeds at birdtables | Feeds at hanging feeders | Uses nest boxes

WESTERN TANAGER

SIZE: 7¼ inches (18 cm)

HABITAT: Coniferous forests

BREEDS: Western North America

WINTERS: Mexico and farther south

FOOD: Insects and berries

7¼ in.
18 cm

Aw As ♂

 KEY: ♂ Male   ♀ Female   Aw Adult winter   As Adult summer

LAZULI BUNTING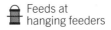

SIZE: 5½ inches (14 cm)

HABITAT: Brushy areas, thickets, backyards

BREEDS: Washington and Montana, south to southern California

WINTERS: Northen Arizona and New Mexico

FOOD: Seeds, insects

As ♂

EVENING GROSBEAK

SIZE: 7–8 inches (18–20 cm)

HABITAT: Conifers, shrubs, gardens

BREEDS: Canada, western and northeastern U.S.

WINTERS: Southward to Mexico

FOOD: Seeds, fruits, berries

As ♂

| **J** Juvenile | Feeds at birdtables | Feeds at hanging feeders | Uses nest boxes |

95

Bird Logbook

This section is a logbook to help you keep track of your explorations and observations of the world of birds. There are pages for every week of the year, offering tips and prompts for recording birdlife and activity in your backyard or locality.

Important things to note include the date of the observations, the time, your location, perhaps the weather, and a list of the birds that you see. For more tips, refer to the section, "Taking Notes and Making Sketches," on pages 50–53.

YEAR ONE

WEEK: STARTING / /

SPECIES	NO.	FOOD TAKEN	TIME	WEATHER

Notes

WEEK:

SPECIES	NO.	FOOD TAKEN	TIME	WEATHER

Notes

BIRD TRACKER TIP

If you see a bird you can't identify, don't jump to conclusions too quickly. Look carefully and check your identification book to see what is likely and what is not: the one that "looks right" at first glance in the pictures might not be found in your area or not at that time of year, so check the text too.

YEAR ONE

WEEK: STARTING / /

SPECIES	NO.	FOOD TAKEN	TIME	WEATHER

Notes

_____

_____

_____

_____

_____

_____

BIRD TRACKER TIP

Feeding birds in winter is good fun, brings them close, and helps the birds survive bad weather. But you should be careful to avoid potential sources of disease and infection: keep the feeders clean, move your feeding station occasionally, and make sure water is not fouled by dirt and droppings. Killing birds by intended kindness is easily avoided.

WEEK:

SPECIES	NO.	FOOD TAKEN	TIME	WEATHER

Notes

YEAR ONE

WEEK: *STARTING   /   /*

SPECIES	NO.	FOOD TAKEN	TIME	WEATHER

Notes

WEEK: *STARTING / /*

SPECIES	NO.	FOOD TAKEN	TIME	WEATHER

Notes

_____

_____

_____

_____

_____

_____

_____

BIRD TRACKER TIP

Try to learn the basic anatomy of your bird (see pages 14–25). This will help you understand the bird's structural differences and its subsequent development. An understanding of the fundamentals, however vague, will enhance your awareness, and may help when you need to communicate with an expert in the field.

WEEK:

STARTING / /

SPECIES	NO.	FOOD TAKEN	TIME	WEATHER

Notes

BIRD TRACKER TIP

Watch birds on the feeders and bird table and how they move around in your garden. They are remarkably quick: see how they take off from a bush, fly to the feeder, and "stop dead" when they reach it—small birds are almost insect-like in their stop-start movements and precision. Just think of the coordination between eyes, wings, and feet required to land on the side of a vertical nut basket.

WEEK:

SPECIES	NO.	FOOD TAKEN	TIME	WEATHER

Notes

YEAR ONE

WEEK:

SPECIES	NO.	FOOD TAKEN	TIME	WEATHER

Notes

WEEK:

SPECIES	NO.	FOOD TAKEN	TIME	WEATHER

Notes

BIRD TRACKER TIP

Birds use all kinds of signals to communicate: at a feeder, you will normally see aggression more than any other interaction. They use their voices but also visual signals, opening their wings and flattening their feathers to look big, wide, and impressive—an unmistakable "get out of my way or else." Such signals are there to avoid physical fighting, but you might see fierce fights over food, nevertheless.

YEAR ONE

WEEK:

STARTING   /   /

SPECIES	NO.	FOOD TAKEN	TIME	WEATHER

Notes

BIRD TRACKER TIP

Keep an eye on the birds feeding outside your window: how many do you see at one time? You might have five or ten chickadees, but during the day, who knows how many will pass by? In many gardens, there might be five or ten times as many as this every day. Such a throughput might not be noticed unless the birds are individually marked with colored bands.

WEEK:

SPECIES	NO.	FOOD TAKEN	TIME	WEATHER

Notes

YEAR ONE

WEEK:

SPECIES	NO.	FOOD TAKEN	TIME	WEATHER

Notes

WEEK:

SPECIES	NO.	FOOD TAKEN	TIME	WEATHER

Notes

BIRD TRACKER TIP

Do all the birds feed in the same way around your feeders and bird table? Almost certainly there will be several different techniques and you can work out what is going on with a bit of close attention. Most birds eat at the table, or the feeder, but others take a nut or seed and fly off with it, to feed undisturbed nearby, or to store the food for later.

WEEK:

STARTING / /

SPECIES	NO.	FOOD TAKEN	TIME	WEATHER

Notes

_____

_____

_____

_____

_____

_____

BIRD TRACKER TIP

Birds must drink and bathe, even in the coldest weather, to keep themselves in good condition. Fresh, clean water is a lifeline for them, but it is difficult to keep it free from ice in winter. A floating ball might sometimes help: a tiny pump could be practical. But normally you just need to top it off regularly, and what you must never do is to use any kind of antifreeze additive.

WEEK:

SPECIES	NO.	FOOD TAKEN	TIME	WEATHER

Notes

YEAR ONE

WEEK: *STARTING* / /

SPECIES	NO.	FOOD TAKEN	TIME	WEATHER

Notes

WEEK:

STARTING   /   /

SPECIES	NO.	FOOD TAKEN	TIME	WEATHER

Notes

BIRD TRACKER TIP

Some birds just don't feed on the food in hanging baskets or feeders and on tables. They just carry on in the same, natural way, feeding in the hedge or shrubs or on the ground underneath. You can help some of these species in bad weather by scattering grated cheese or other foods on the ground, under hedges and in flowerbeds, or even smearing fat in the crevices in tree bark.

YEAR ONE

WEEK:

SPECIES	NO.	FOOD TAKEN	TIME	WEATHER

Notes

BIRD TRACKER TIP

In spring, birds are looking for good nesting material. You can help them, and have a bit of extra amusement, if you put out bundles of straw, bits of wool, or even clumps of hair—from the dog, or even your own! A solid-wire basket is best to hold such material—avoid soft plastic mesh, and avoid long strands of cotton or other fiber that might entangle birds' legs.

WEEK:

STARTING   /   /

SPECIES	NO.	FOOD TAKEN	TIME	WEATHER

Notes

YEAR ONE

WEEK:

SPECIES	NO.	FOOD TAKEN	TIME	WEATHER

Notes

WEEK:

SPECIES	NO.	FOOD TAKEN	TIME	WEATHER

Notes

_____

_____

_____

_____

_____

_____

BIRD TRACKER TIP

Many birds nest close to people. They are normally not especially prone to suffer from disturbance, but all the same, if you suspect a nest, it is best to avoid it if you can. It is usually safe to take a peek, but to do so at the wrong time might cause a bird to desert its eggs or chicks a day or two before they are ready.

YEAR ONE

WEEK:

SPECIES	NO.	FOOD TAKEN	TIME	WEATHER

Notes

_____

_____

_____

_____

_____

_____

BIRD TRACKER TIP

Look out for signs of woodpeckers on tree bark. Sometimes they chip away bits of bark or decaying wood to get at beetle larvae; sometimes they will drill small, round holes into smooth bark so that sticky sap oozes out; they return to feed on the sap and any insects that happen to get stuck in it. In North America, a small group of woodpeckers that habitually do this are known as sapsuckers.

WEEK:

SPECIES	NO.	FOOD TAKEN	TIME	WEATHER

Notes

YEAR ONE

WEEK:

SPECIES	NO.	FOOD TAKEN	TIME	WEATHER

Notes

WEEK:

SPECIES	NO.	FOOD TAKEN	TIME	WEATHER

Notes

BIRD TRACKER TIP

Try to learn a bird song or call each week. It is always best to match the sound
with the bird by following up the call and seeing exactly what is making it: then
fix it in your mind. You might make up a word or phrase that is suggestive
of the rhythm or pattern of the sound that will help you to recall it later.

YEAR ONE

WEEK:

SPECIES	NO.	FOOD TAKEN	TIME	WEATHER

Notes

_____

_____

_____

_____

_____

_____

_____

BIRD TRACKER TIP

Check your binoculars now and then: look at the individual eyepiece adjustment to make sure it has not moved from your ideal position. We all have a different balance between our eyes, so it is important to get the adjustment right and keep it right. Clean the lenses, blowing away dust or crumbs before you do any wiping with a soft, clean cloth.

WEEK:

SPECIES	NO.	FOOD TAKEN	TIME	WEATHER

Notes

YEAR ONE

WEEK: *STARTING   /   /*

SPECIES	NO.	FOOD TAKEN	TIME	WEATHER

Notes

WEEK:

SPECIES	NO.	FOOD TAKEN	TIME	WEATHER

Notes

BIRD TRACKER TIP

Try photographing your local birds: they are as challenging, and rewarding, as any other wildlife subject. Try to get some action and atmosphere into your pictures. The perfect, static, beautifully lit, and sharply focused portrait is something to aim for, but everyone does that—why not be different?

WEEK:

SPECIES	NO.	FOOD TAKEN	TIME	WEATHER

Notes

BIRD TRACKER TIP

If you aren't too squeamish, offering live food can get your birds hand-tame quicker than anything else. Put a few mealworms in a small container on a table or the ground, then get the birds used to you being around. When they have become accustomed to you, hold out a hand close to them as they feed—then put the food on your palm and see if they will take it.

WEEK:

SPECIES	NO.	FOOD TAKEN	TIME	WEATHER

Notes

YEAR ONE

WEEK:

SPECIES	NO.	FOOD TAKEN	TIME	WEATHER

Notes

WEEK:

SPECIES	NO.	FOOD TAKEN	TIME	WEATHER

Notes

BIRD TRACKER TIP

If you have any ability at all, sketching birds is really rewarding: but it is not easy. They do not stay still for long. Just sketch while you can, fast and loose, and if the bird flies off, start on the next. Just keep doing it until you get more confident and begin to capture the birds' expressions and character.

YEAR ONE

WEEK: *STARTING   /   /*

SPECIES	NO.	FOOD TAKEN	TIME	WEATHER

Notes

BIRD TRACKER TIP

If you draw birds, try to draw what you see in front of you rather than what you know about from other sources. Details of the feet, all the outlines of the feathers—these are things that you may have seen in a book, but most of the time you can't see them when observing birds, so why draw them? Get real: put in what you see and nothing else and your pictures will come alive.

WEEK:

SPECIES	NO.	FOOD TAKEN	TIME	WEATHER

Notes

YEAR ONE

WEEK:

SPECIES	NO.	FOOD TAKEN	TIME	WEATHER

Notes

WEEK:

STARTING   /   /

SPECIES	NO.	FOOD TAKEN	TIME	WEATHER

Notes

BIRD TRACKER TIP

If you put out peanuts to feed birds, keep them chopped up in tiny pieces, or put them in a small-mesh basket so that the nuts last longer. This would also prevent the occasional rare incidence of a baby bird choking on a large peanut given by its parent.

YEAR ONE

WEEK:

STARTING   /   /

SPECIES	NO.	FOOD TAKEN	TIME	WEATHER

Notes

_____

_____

_____

_____

_____

_____

_____

BIRD TRACKER TIP

A nest box can work well in a garden, but do not be tempted to put out too many for the size of the garden. Most birds are territorial and too many boxes in a small area are likely to cause birds to spend all day fighting each other instead of rearing their young. Keep boxes away from feeders, too: nesting birds need peace and seclusion.

WEEK: _STARTING_ / /

SPECIES	NO.	FOOD TAKEN	TIME	WEATHER

Notes

YEAR ONE

WEEK: *STARTING   /   /*

SPECIES	NO.	FOOD TAKEN	TIME	WEATHER

Notes

WEEK: *STARTING   /   /*

SPECIES	NO.	FOOD TAKEN	TIME	WEATHER

Notes

_____

_____

_____

_____

_____

BIRD TRACKER TIP

If you are a gardener, consider the wildlife when you are planting up a new border or patch of ground. Search the Internet for details of native plants in your area and match what you use to your soil; remember that flowers with an abundance of nectar are good for insects, which are good for birds. In North America, flowers with abundant nectar are also good for hummingbirds.

YEAR ONE

WEEK: _____   *STARTING   /   /*

SPECIES	NO.	FOOD TAKEN	TIME	WEATHER

Notes

_____

_____

_____

_____

_____

_____

_____

BIRD TRACKER TIP

If you have space, a garden pond is a great idea. It should be situated away from hedges and trees that would fill it with leaves in fall. Dig twice the depth you need, then add a piece of old carpet, a rubber or plastic pond liner, and some soil. Be careful not to use non-native water plants, which could germinate elsewhere and cause havoc in local streams.

WEEK:

STARTING / /

SPECIES	NO.	FOOD TAKEN	TIME	WEATHER

Notes

_____
_____
_____
_____
_____
_____
_____
_____
_____
_____
_____
_____
_____
_____
_____

YEAR ONE

WEEK:

STARTING   /   /

SPECIES	NO.	FOOD TAKEN	TIME	WEATHER

Notes

WEEK:

SPECIES	NO.	FOOD TAKEN	TIME	WEATHER

Notes

BIRD TRACKER TIP

If you have a pond, make sure at least part of its edge is shallow and shelves smoothly up to dry ground. That way, even the smaller birds can walk in and bathe or drink, without having to dive into deep water. A few well-placed stones around the edge will help to keep open spaces between clumps of vegetation.

YEAR ONE

WEEK:

SPECIES	NO.	FOOD TAKEN	TIME	WEATHER

Notes

_____
_____
_____
_____
_____
_____
_____

BIRD TRACKER TIP

If you visit the seashore or marsh, make sure you check the times of the tides, both for your own safety and as a guide to bird activity. Shore birds are often resting in a protected place at high tide, while at low tide they come out in numbers to feed in exposed areas.

WEEK:

SPECIES	NO.	FOOD TAKEN	TIME	WEATHER

Notes

YEAR ONE

WEEK: *STARTING   /   /*

SPECIES	NO.	FOOD TAKEN	TIME	WEATHER

Notes

WEEK:

SPECIES	NO.	FOOD TAKEN	TIME	WEATHER

Notes

BIRD TRACKER TIP

A vacation is a great time to see new birds in different habitats. Get used to using a topographic map: good maps are mines of information of value to the nature-watcher. Find appropriate habitats, such as woods, streams, lakes, and marshes, and good vantage points that will give you an extensive view with the sun behind you at the right time of day.

YEAR ONE

WEEK:

SPECIES	NO.	FOOD TAKEN	TIME	WEATHER

Notes

BIRD TRACKER TIP

There's no harm looking at the bird-watching year and trying to plan what to do next year, to see the highlights over again or to get to see new ones. A year goes quickly and fitting in all the birds you would like to see is barely possible. Identify a few gaps now, and think about how, when, and where you might fill them in the year to come.

WEEK:

STARTING   /   /

SPECIES	NO.	FOOD TAKEN	TIME	WEATHER

Notes

YEAR TWO

WEEK:

SPECIES	NO.	FOOD TAKEN	TIME	WEATHER

Notes

WEEK:

SPECIES	NO.	FOOD TAKEN	TIME	WEATHER

Notes

BIRD TRACKER TIP

If only once, you should make a point of getting up very early and going to a local woodland to listen to the dawn chorus: it is remarkable and moving, something that can hardly be described but has to be experienced firsthand for the full, spectacular effect to be properly appreciated.

YEAR TWO

WEEK:

SPECIES	NO.	FOOD TAKEN	TIME	WEATHER

Notes

BIRD TRACKER TIP

Water is always a bonus on any bird-watching trip: try to find a good lake, reservoir, or flooded pit and you will undoubtedly enjoy a great variety of species, some of them probably best appreciated at close range. A quiet summer evening is a lovely time to visit such a place.

WEEK:

SPECIES	NO.	FOOD TAKEN	TIME	WEATHER

Notes

YEAR TWO

WEEK: *STARTING   /   /*

SPECIES	NO.	FOOD TAKEN	TIME	WEATHER

Notes

_____
_____
_____
_____
_____
_____
_____
_____
_____
_____
_____
_____
_____
_____

WEEK: _STARTING   /   /_

SPECIES	NO.	FOOD TAKEN	TIME	WEATHER

Notes

_____

_____

_____

_____

_____

_____

_____

BIRD TRACKER TIP

Some migrants begin to move south surprisingly early, in late summer or early fall. It is always easy to take note of the first migrant you see in spring but much less so to notice the last of any species that you see in the year. Keeping notes of such observations will help you anticipate future movements of various species.

YEAR TWO

WEEK: *STARTING   /   /*

SPECIES	NO.	FOOD TAKEN	TIME	WEATHER

Notes

BIRD TRACKER TIP

Migrating wading birds can turn up at the edge of a pool almost anywhere in spring and fall. To see them locally you may need to return to a suitable watery habitat many times, as some may stay for days but others may be present for only a few hours.

WEEK:

SPECIES	NO.	FOOD TAKEN	TIME	WEATHER

Notes

YEAR TWO

WEEK:

SPECIES	NO.	FOOD TAKEN	TIME	WEATHER

Notes

WEEK:

SPECIES	NO.	FOOD TAKEN	TIME	WEATHER

Notes

BIRD TRACKER TIP

Bird guides on video and online show aspects of birds' movement and song
and such guides can also carry large amounts of other information. Use these
sources in combination with the precision and portability of your pocket
guidebook, which is unbeatable for quick reference while you are watching birds.

YEAR TWO

WEEK:

STARTING   /   /

SPECIES	NO.	FOOD TAKEN	TIME	WEATHER

Notes

_____

_____

_____

_____

_____

_____

_____

BIRD TRACKER TIP

Listen for migrants passing overhead at night. There is no point spending hours outside doing this, but if you are outside after dark, be aware that some birds might be passing overhead and listen for them. It is surprising what unexpected calls might be heard even over urban areas during migration periods.

WEEK:

SPECIES	NO.	FOOD TAKEN	TIME	WEATHER

Notes

YEAR TWO

WEEK:

STARTING   /   /

SPECIES	NO.	FOOD TAKEN	TIME	WEATHER

Notes

WEEK:

SPECIES	NO.	FOOD TAKEN	TIME	WEATHER

Notes

BIRD TRACKER TIP

Television weather forecasts are an excellent resource for bird-watchers, especially detailed forecasts with proper synoptic charts. It can be possible to predict "good days" when it is worth traveling to the coast or a migration watch point—especially if there is a clear night with tail winds, which is good for migration, followed by poor weather, cloud, or fog, which forces birds to land.

YEAR TWO

WEEK: *STARTING   /   /*

SPECIES	NO.	FOOD TAKEN	TIME	WEATHER

Notes

BIRD TRACKER TIP

On a bright, clear day during spring or fall migration, it may be worthwhile
getting up on a hillside with a good view and watching for birds of prey on the
move. They often tend to follow long ridges or chains of hills, using updrafts
to help keep them aloft with little effort. If you can find such a place, you
might see some interesting travelers.

WEEK:

SPECIES	NO.	FOOD TAKEN	TIME	WEATHER

Notes

YEAR TWO

WEEK: _STARTING_   /   /

SPECIES	NO.	FOOD TAKEN	TIME	WEATHER

Notes

_____
_____
_____
_____
_____
_____
_____
_____
_____
_____
_____
_____
_____
_____
_____
_____

WEEK:

SPECIES	NO.	FOOD TAKEN	TIME	WEATHER

Notes

BIRD TRACKER TIP

Wet weather can be good for birds but is not so good for watching them. A blind gives good shelter, but even good, breathable, waterproof fabrics may not make a day out in the rain especially comfortable. Don't neglect the old-fashioned umbrella—with a hooked handle to curl under one arm, it can be a great help.

YEAR TWO

WEEK:

STARTING   /   /

SPECIES	NO.	FOOD TAKEN	TIME	WEATHER

Notes

BIRD TRACKER TIP

Migrant birds need plenty of nutritious food to give them a high-energy intake for the least amount of effort. Bushes with juicy, sugar-rich berries strongly attract some species of migrant warblers, concentrating local populations into the best feeding areas—if you can find these areas, you will see a lot of birds.

WEEK:

SPECIES	NO.	FOOD TAKEN	TIME	WEATHER

Notes

YEAR TWO

WEEK: *STARTING   /   /*

SPECIES	NO.	FOOD TAKEN	TIME	WEATHER

Notes

WEEK:

STARTING   /   /

SPECIES	NO.	FOOD TAKEN	TIME	WEATHER

Notes

BIRD TRACKER TIP

Be aware of your behavior and surroundings when you are out in the countryside—if you want to see birds, you must keep quiet. Approach the edges of habitats, streams, and rivers with special care, looking both ways to try to see birds before they see you.

YEAR TWO

WEEK: _____

STARTING   /   /

SPECIES	NO.	FOOD TAKEN	TIME	WEATHER

Notes

_____

_____

_____

_____

_____

_____

_____

BIRD TRACKER TIP

Always be prepared to learn from other people's experiences. There is a wealth of information now, in books and magazines, on Web sites—including forums and question pages—and from various clubs and societies. "Look and learn" is the motto. Contribute, too, as you may still see something that no one else has.

WEEK:

SPECIES	NO.	FOOD TAKEN	TIME	WEATHER

Notes

YEAR TWO

WEEK: *STARTING   /   /*

SPECIES	NO.	FOOD TAKEN	TIME	WEATHER

Notes

_____
_____
_____
_____
_____
_____
_____
_____
_____
_____
_____
_____
_____
_____

WEEK:

SPECIES	NO.	FOOD TAKEN	TIME	WEATHER

Notes

BIRD TRACKER TIP

Look at birds' plumages in fall and see how, if at all, their colors and patterns
differ between seasons, with age, or between male and female. With small birds up
close—as at the kitchen window—or big birds in the countryside, it is fascinating to
see the progress of their fall molt, as gaps appear and fill again in their feathers.

YEAR TWO

WEEK:

STARTING / /

SPECIES	NO.	FOOD TAKEN	TIME	WEATHER

Notes

BIRD TRACKER TIP

When in the countryside, all bird-watchers should respect the welfare of birds, domestic and other animals, and the rights of other people and landowners. Observe practices such as not leaving litter, closing gates, and clearing up after your dog, if you have one.

WEEK:

SPECIES	NO.	FOOD TAKEN	TIME	WEATHER

Notes

WEEK: *STARTING   /   /*

SPECIES	NO.	FOOD TAKEN	TIME	WEATHER

Notes

WEEK:

SPECIES	NO.	FOOD TAKEN	TIME	WEATHER

Notes

BIRD TRACKER TIP

Cats can be a menace in a bird-watcher's garden. Various pellets and liquid deterrents may help keep dogs and cats away, but many are determined and not so easily put off. An electronic cat deterrent, which gives an ultrasound bleep when triggered, can be more effective, and does no harm to the neighbors' pets.

YEAR TWO

WEEK: *STARTING   /   /*

SPECIES	NO.	FOOD TAKEN	TIME	WEATHER

Notes

BIRD TRACKER TIP

You may see birds that simply don't look like anything you have in your books. These may be "escapes," such as parrots, finches, weavers, and other cage birds, that have been released or have escaped and managed to survive for a time. Others are simply aberrant in some way, such as albinos, piebald leucistic birds (which have an excess of white), or abnormally pale individuals with reduced feather pigment.

WEEK:

SPECIES	NO.	FOOD TAKEN	TIME	WEATHER

Notes

YEAR TWO

WEEK: _____   *STARTING* / /

SPECIES	NO.	FOOD TAKEN	TIME	WEATHER

Notes

WEEK:

SPECIES	NO.	FOOD TAKEN	TIME	WEATHER

Notes

BIRD TRACKER TIP

If you want to watch birds with other people, try visiting a nature reserve and joining an organized walk with the guide. You could also join a local bird club or group and go on their trips.

YEAR TWO

WEEK:

STARTING   /   /

SPECIES	NO.	FOOD TAKEN	TIME	WEATHER

Notes

_____

_____

_____

_____

_____

_____

_____

BIRD TRACKER TIP

For a good idea of how different species of birds get along in the same area or habitat, go to a tidal beach and watch the shorebirds. Wading birds with different lengths of leg and bill, or different bill shapes, feed on different foods, some in water, some in mud, some on sand, some picking from the surface, some probing more deeply.

WEEK:

SPECIES	NO.	FOOD TAKEN	TIME	WEATHER

Notes

YEAR TWO

WEEK: *STARTING* / /

SPECIES	NO.	FOOD TAKEN	TIME	WEATHER

Notes

WEEK:

SPECIES	NO.	FOOD TAKEN	TIME	WEATHER

Notes

BIRD TRACKER TIP

Like shorebirds, ducks have a great variety of behavior: some dive, feeding on animal matter or plant matter; others graze on dry land; others dabble in the shallow water at the edge of a lake. Some species seem to be constantly active by day, others sleep, feeding mostly at night. They are worth watching to try to gain a better understanding of how variety can be supported in a small area.

YEAR TWO

WEEK:

SPECIES	NO.	FOOD TAKEN	TIME	WEATHER

Notes

BIRD TRACKER TIP

Garden birds need food and feeding opportunities, and supplying artificial foods, or natural foods in artificial feeding sites, is not always enough. A mixture of shrubs and trees, flowerbeds and lawns, bare earth and vegetation, will ensure a mixture of birds; shingle, concrete, decking, and slabs will usually do the opposite.

WEEK:

SPECIES	NO.	FOOD TAKEN	TIME	WEATHER

Notes

YEAR TWO

WEEK: STARTING / /

SPECIES	NO.	FOOD TAKEN	TIME	WEATHER

Notes

WEEK:

STARTING / /

SPECIES	NO.	FOOD TAKEN	TIME	WEATHER

Notes

BIRD TRACKER TIP

How many birds hang around your house or the office building, or sit around on roofs in town, but hardly seem to come to the ground? It is interesting to watch town and suburban birds and see just where they feed and what they do all day. Gulls and crows might appear to stand on the roof or the television antenna all day long, but they obviously do come down to eat somewhere, sometime—see if you can find out more.

YEAR TWO

WEEK:

SPECIES	NO.	FOOD TAKEN	TIME	WEATHER

Notes

BIRD TRACKER TIP

Listen for very quiet bird song. Sometimes a bird will sing—as if "singing to itself" or "daydreaming"—so quietly that you have to be very close to hear it. The song may be like normal song or quite different. This type of singing is known as subsong: it occurs often in young birds learning their genetically programmed species repertoire or in adult birds early in the breeding season.

WEEK:

STARTING   /   /

SPECIES	NO.	FOOD TAKEN	TIME	WEATHER

Notes

YEAR TWO

WEEK:

SPECIES	NO.	FOOD TAKEN	TIME	WEATHER

Notes

WEEK:

STARTING / /

SPECIES	NO.	FOOD TAKEN	TIME	WEATHER

Notes

BIRD TRACKER TIP

If you get a fall of snow, look for tracks and other signs of birds and other
wildlife. Birds may leave telltale marks, such as the sweep of wingtips
made by a hawk diving on its prey, or the giant footprints of a heron coming
unsuspected to your garden pond at dawn. Mud is a fair substitute.

YEAR TWO

WEEK:

SPECIES	NO.	FOOD TAKEN	TIME	WEATHER

Notes

BIRD TRACKER TIP

If you find feathers, try to identify which species and which part of the body
they came from. The distinctive stiff quills of flight and tail feathers are
quite unlike the softer contour feathers of the body. A shed feather reveals
unexpected patterns, too, as spots and bars become apparent on the inner
edge that are usually hidden on the living bird.

WEEK:

SPECIES	NO.	FOOD TAKEN	TIME	WEATHER

Notes

YEAR TWO

WEEK: *STARTING   /   /*

SPECIES	NO.	FOOD TAKEN	TIME	WEATHER

Notes

WEEK:

SPECIES	NO.	FOOD TAKEN	TIME	WEATHER

Notes

BIRD TRACKER TIP

Baby birds that seem to be "abandoned" are best left alone, as their parents will likely be nearby. If a bird is sick or injured, do not try to look after it. Most towns and cities in the U.S. have bird rehabilitators nearby and birders should make a point of learning who they are. Paying them a visit may allow you to see birds up close that are sometimes difficult to see.

YEAR TWO

WEEK:

STARTING   /   /

SPECIES	NO.	FOOD TAKEN	TIME	WEATHER

Notes

BIRD TRACKER TIP

As you build up bird diaries or notebooks, it is useful to compare a past year's notes
with the current year to see whether the bird life in your area has changed in any way.
Are there more or fewer birds of certain species? Did they come and go earlier or later
this year? Half the fun of keeping notes is to compare them over time.

WEEK:

STARTING   /   /

SPECIES	NO.	FOOD TAKEN	TIME	WEATHER

Notes

Glossary

ABRASION Wear and tear on feathers, often removing paler spots and fringes and fading darker colors.

ALBINISM A lack of pigment: true albinos are white with pink eyes, but most "white" birds are partial albinos, or albinistic, with patches of white and normal eye colors.

AXILLARIES The feathers under the base of the wing, in the "wingpit." Also known as axillars.

BAND A metal band placed around a bird's leg, with an individual number; when the bird is caught or found dead, its movements can be traced. Also known as a ring.

BASTARD WING A tuft of feathers on the "thumb," halfway along the leading edge of the wing, which can be raised or lowered to control airflow in flight. Also called the alula.

BEAK Synonymous with bill; the two jaws and their horny covering.

BINOCULAR VISION The ability to see an area with both eyes; birds such as owls have forward-facing eyes, giving the greatest extent of binocular vision.

BIRD OF PREY Usually refers to daytime birds of prey, including eagles, vultures, hawks, falcons, harriers, and kites; may be used to include owls. Also called "raptors," or raptorial birds.

BLIND A small shelter from which to observe birds while remaining hidden from view. Also known as a hide.

BROOD A set of young birds hatched from one clutch of eggs.

CALL NOTE A vocalization, usually characteristic of the species, made to maintain contact, warn of danger, or for other specific purposes.

CAP A patch of color on the top of a bird's head, usually on the feathers of the forehead and crown.

CARPAL JOINT The bend of the wing, at the "wrist."

CHICK A young bird before it is able to fly.

CLUTCH A set of eggs laid and incubated together in the nest; if these are lost, a replacement clutch may be laid; some species have several clutches during the course of one breeding season, others ("single brooded") have only one.

COLONY A group of nests close together, often on the ground (e.g., gulls and terns) or in trees (e.g., herons).

COLOR RING OR BAND A plastic or metal band placed on a bird's leg; a combination of colors or numbers on the band allow individual recognition without having to capture the bird.

CORVID A bird of the crow family or corvidae.

COURTSHIP Usually ritualized behavior, male and female together forming a pair bond before breeding.

CRYPTIC Describes coloration that gives a bird camouflage or makes it harder to see.

DAWN CHORUS The loud chorus of bird song heard in spring from just before dawn, especially in woodland.

DISPLAY A form of ritualized behavior with a specific function, for example in courtship, or in distracting potential predators.

DISTRIBUTION The geographical range of a species, often split into breeding range, wintering range, and areas in which it may be seen on migration.

DRAKE A male duck (females are then "ducks").

DRUMMING The sound made in spring by a woodpecker vibrating its bill against a branch; also made by a snipe diving through the air with outer tail feathers extended and vibrating.

DUSTING "Bathing" in loose, dry sand, dust, or soil to help remove parasites from feathers.

ECLIPSE A dull plumage worn by male ducks and geese in summer.

EXTINCT Describes a species no longer living anywhere on Earth; if a species has disappeared from a country or region, but is still found elsewhere, it is properly described as having been "extirpated" from that area.

FALL A sudden large arrival of migrant birds, especially when caused by bad weather on the coast.

FERAL Describes a bird or species that has escaped from captivity to live wild.

FIELD "In the field" means "in the wild" or out of doors (as opposed to being captive, or held "in the hand").

FIELD GUIDE An identification guide to birds as they are seen wild and free.

FIELD OF VIEW The extent of the area that can be seen through a telescope or binoculars at a given distance, expressed in degrees (angular field of view) or distance (linear field of view); higher magnification typically results in a smaller field of view.

FLEDGLING A young bird that has just learned to fly and has its first covering of feathers.

FLOCK A group of birds behaving in some sort of unison: tight flocks (e.g., starlings in flight) are obvious, but loose, feeding flocks of birds in woodland may be less so.

GAME BIRD Usually used to describe one of the pheasant, partridge, grouse, or quail families—other birds commonly shot for sport include ducks and geese ("waterfowl").

GENUS A category in classification above species, indicating close relationships. Appears as the first word in a two- or three-word scientific name (e.g., *homo* in *homo sapiens*, or *falco* in *falco peregrinus*). Plural is "genera."

GORGET Band of color or pattern, such as streaks, around the bird's upper breast.

HABITAT The environment that a species requires for survival. Its characteristics include shelter, water, food, feeding areas, nest sites, and roosting sites. More loosely described in such terms as "lowland heath" or "deciduous woodland"; also used for particular times of year or types of behavior, e.g., muddy estuary, open sea, ploughed fields.

HEN A female bird.

IMMATURE Describes a bird not yet old enough to breed or have full adult plumage colors.

INCUBATION Maintenance of proper temperature of the egg to allow development of the embryo.

JIZZ A kind of indefinable quality that gives a species a character of its own, combining shape, color, and—especially—actions.

JUVENILE The young bird in its first full plumage. Also known as juvenal in the U.S.

LOAFING Sitting or standing, often in groups, apparently doing little or nothing. Gulls, for example, "loaf" for hours at a time.

MANDIBLE The jaw and its horny sheath; upper and lower mandibles together form the beak or bill.

MEASUREMENTS The size of a bird is usually indicated by the length from bill tip to tail tip on a bird laid out on a flat surface. In reality, the apparent "size" depends as much on shape and bulk as on simple length.

MIGRATION A regular, seasonal movement of birds from one region or continent to another, between alternate areas occupied at different times of year.

MOLT The replacement of a bird's feathers, in a regular sequence characteristic of each species. There may be a complete molt or a partial molt depending on the season.

NEST A receptacle built to take a clutch of eggs and, in many species, the young birds before they are able to fly; eggs may also be laid on a bare ledge or on the ground, with no nest structure being made.

NOCTURNAL Active at night.

NUMBERS Bird populations vary hugely from season to season, so are best described in terms of a particular measure that is easily repeated, usually "breeding pairs." In the case of large, more easily counted birds, such as ducks and geese, the measure is the total number of individuals at a certain season.

ORNITHOLOGY The study of birds: usually refers to scientific study of biology and ecology, while the hobby of watching birds is known simply as bird-watching or birding.

PASSAGE MIGRANT A species or bird seen in some intermediate area during its migration from summer to winter quarters (or vice versa).

PASSERINE A "perching bird."

PLUMAGE A covering of feathers; also often used to describe the overall colors and patterns of the feathers, defining a bird's appearance according to age, sex, and season.

PREENING Care of the feathers, especially using the bill to "zip" the structures back into place.

RACE A recognizable geographical group, or subspecies, within a species. Often there is no obvious border between groups, which blend (in a "cline") from one extreme to another. There may be more distinctive differences between isolated areas, such as islands, in which case the decision whether there are races, or separate species, can be difficult.

RARITY An individual bird in an area where it is not normally seen, or is seen in only very small numbers. A species with a small world population is "rare."

ROOST To sleep; also the area where birds sleep.

SCAPULARS A bunch of feathers on the shoulder.

SEABIRD A species that comes to land to nest, but otherwise lives at sea and is not normally seen inland.

SOARING Flight, often at a high level, in which the wings are held almost still, using air currents for lift and propulsion.

SONG A vocalization with a specific purpose and usually distinctive for each species: in particular, advertising the presence of a bird on its territory.

SPECIES A group, or groups, of individuals that can produce fertile young. Different species rarely interbreed naturally; if they do so, infertile hybrid offspring are produced.

TERRITORY An area defended for exclusive use by an individual bird or a family. Both breeding and winter-feeding territories may be defended.

TWITCHER A bird-watcher temporarily engaged in "twitching" (hearing of the presence of an individual rare bird and traveling with the intention of seeing it). Not then, despite the media's frequent incorrect usage, a bird-watcher, but a particular kind of bird-watcher.

WADER A plover, sandpiper, curlew, or related species; in North America, usually called a "shorebird." Since some do not wade and some do not live on the shore, neither word is entirely satisfactory.

WATERFOWL Ducks, geese, and swans. Also known as wildfowl.

Index

Acknowledgments

Photography: Andrew Syred/Science Photo Library (p16).
Illustrations: Norman Arlott, Dianne Breeze, Keith Brewer, Hilary Burn, Chris Christoforou, Richard Draper, Malcolm Ellis, Mark Franklin, Robert Gillmor, Tony Graham, Vana Haggerty, Peter Hayman, Gary Hincks, John Hutchinson, Aziz Khan, Colin Newman, Denys Ovenden, David Quinn, Andrew Robinson, Chris Rose, Ed Stuart, David Thelwell, Gill Tomblin, Owen Williams, Ann Winterbotham, Ken Wood, and Michael Woods.

FEB - 9 2023